HORSTMANN'S:
THE ENTERPRISE OF MILITARY EQUIPAGE

by
Bruce S. Bazelon

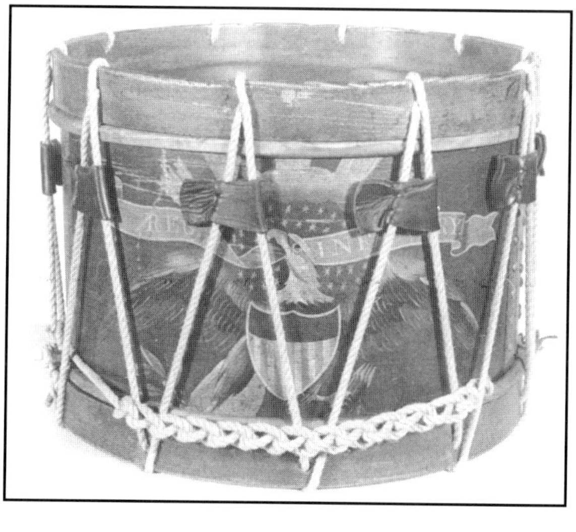

Dedication:
To the memory of my father, Dr. A. Lloyd Bazelon

For information or additional copies, please contact
Bruce S. Bazelon
319 North 24th Street
Camp Hill, PA 17011

Typeset by REF Typesetting & Publishing, Inc.
9400 Fairview Ave., Manassas, VA 20110

Acknowledgements

I am greatly indebted to the interest of many individuals who pointed the way to sources which might otherwise have gone undetected. In particular, the staff of the Pennsylvania State Library put up with my nosing through the stacks, the genealogical collections on microfilm, the newspapers, and many other sources for some years. They also obtained many otherwise unreachable materials for this project through interlibrary loan. They certainly deserve many thanks. The Historical Society of Pennsylvania and its Registrar, Janice Dockery, helped greatly by bringing forth and microfilming their own holdings for this project. The Philadelphia Free Library made available the Maxman insurance maps, reprinted in this work. The staff of Eleutherian Mills Library also made accessible their copies of the R. G. Dun and Company Collection, as well as rare business histories from the Library's extensive holdings. The library of the National Museum of American History, the New-York Historical Society, the Library Company of Philadelphia, the Urban Archives of Temple University, and the Philadelphia City Archives also have been consulted for city directories, will registers, and other important source materials. Dr. Plowman of the Philadelphia Branch of the National Archives was also most kind in making available sources such as the Horstmann bankruptcy papers. The National Archives in Washington, DC, yielded the original contract ledgers documenting the Horstmann firm's sales to the U. S. Government. The Lutheran Church Archives, located in the Lutheran Theological Seminary in Germantown Avenue, while not producing specific materials on Horstmann's church membership, provided valuable contextual sources. I am indebted to Mr. Donald Kloster of the National Museum of American History who contributed important interpretations for the purposes of this book.

A number of collectors were most generous in allowing me to photograph surviving Horstmann artifacts. They include Joseph Marino, William McGuinn, Jackie Eubanks, Stuart Mowbray, Craig Nannos, Clarence Runtsch, George Juno and Russ Pritchard of the American Ordnance Preservation Association, Richard Johnson, William Synamon and Kevin Hoffman. A special thanks is extended to Mr. Norm Flayderman for his permission to use the Widmann/Horstmann design book and other materials in his collection. The staff of the State Museum of Pennsylvania and Mr. Donald Kloster of the Smithsonian Institution have again been most helpful in pointing out Horstmann materials in their collections.

This work owes much to J. Duncan Campbell whose own interest in military manufactures provided a point of departure for my own work. A special thanks is due Dr. Caroline Stuckert for her assistance with the research of this work and for helping to craft its development. The author also wishes to acknowledge the efforts of the late Alice Hemenway who undertook the prooofreading of this work only a few months before her untimely death. Additional thanks to Duncan Campbell for his assistance in reading this manuscript and for offering many useful suggestions and insights and to Barbara Bazelon for her encouragement and her willingness to review this work as it developed.

Cut velvet portrait of Napoleon believed to have been made by W. H. Horstmann in Paris about 1810. One Hundred Years, *p. 13.*

Sources

Great men and large enterprises are comparatively easy to write about because their contemporaries and immediate successors saw value in preserving some record of them. Ordinary individuals and the businesses they create usually disappear, leaving only a faint trace unless some extraordinary accident has happened to leave some record.

The Horstmann family and the firm they created were just large enough and well enough known that some of their original records were preserved and indeed some have been. The large stacks of papers sold at the company's bankruptcy sale in 1948 have disappeared. However, through the foresight of Sigmund Horstmann's daughter, Elizabeth J. Lippincott, some Horstmann records have been included with the Coles Lippincott Horstmann family papers at the Historical Society of Pennsylvania. These records include:

- a scrapbook covering the firm's history from approximately 1850 to 1888 with references to earlier events
- a book recording machine purchases from 1842 to 1870
- cash balance sheets from the 1850s and '60s
- ribbon sample books 1847-76
- an employee register c. 1870
- a cash book dealing with some of the firm's sales 1858-1872.

The survival of these meager materials make the Horstmann firm the best documented of the 19th century military goods houses. In the military ornament trade only the records of the button making firms in Waterbury, Scovill, Waterbury Button Company and Steele and Johnson have been better preserved than the surviving Horstmann records. Records of several of the large arms makers have survived. These document the business practices of the mid-19th century weapons manufacturing concerns. The papers of the Colt Patent Firearms Company are in Hartford, and the records of the Winchester Company, together with those of H. K. White, a gun dealer in New York, are now at the Buffalo Bill Historical Center in Cody, Wyoming. Of these sources, only the Scovill papers add substantive material to the present study of Horstmann.

In addition to these primary materials, there are a number of contemporary secondary records of great value. The records of R. G. Dun and Company, now Dun and Bradstreet, are in the Baker Library at Harvard University. Unfortunately, for a successful firm like that of the Horstmanns, business ratings approach hagiography and do not reveal a great deal about the firm. The Census Schedule E documents American manufacturers in 1820, 1850, 1860, 1870, and 1880. Of these, the Horstmann firm is not found in the 1820 census, but a remarkable description of his direct competitor, Henry Korn, also of Philadelphia, is cited in its entirety.

The most important single source is a centennial history produced by Horstmann in 1916, entitled, *One Hundred Years 1816 to 1916, the Chronicles of an Old Business House in the City of Philadelphia.* While the author is not cited, the work was clearly inspired by Walter Horstmann (1857-1922), the grandson of the firm's founder, who was obviously interested in his family's history. In 1916 Walter Horstmann was President of his grandfather's company. Good descriptions of the Horstmann business are also found in Edwin T. Freedley, *Philadelphia and its Manufactures* 1858 and 1859 editions; *The Encyclopedia of Pennsylvania Biography*, vol. XIV, 1923; J. Leander Bishop, *A History of American Manufactures from 1608 to 1860* (1868); Charles Morris, *Makers of Philadelphia* (1894); R. A. Smith, *Philadelphia as it is, in 1852* (1852); and *Manufacturies and Manufacturers of Pennsylvania of the Nineteenth Century* (1875). In addition, bits and pieces appear in a number of other secondary works about Philadelphia history.

Other useful sources are the cemetery records of St. Michael's Lutheran Church in Germantown, as compiled by John H. Heffner (1994), and those of Laurel Hill Cemetery in Philadelphia. These cemetery records yield important life and death dates as well confirming other genealogical data. For the later history of the firm, the bankruptcy dockets of the Horstmann Uniform Company, a direct descendent of the original firm, are now in the Philadelphia Branch of the National Archives. These records show the twentieth century history of the Horstmann concern from its sale in 1923 to its final demise in 1948. These records are supplemented by the *Evening Bulletin* clippings files now in the

Philadelphia Urban Archives at Temple University.

The Horstmann firm is the best documented of American military goods companies in the 19th century. The survival of at least some firm records is a tribute to the size and importance of the Horstmann firm. It was noticed by its contemporaries. Family members took at least some care to document the firm because they believed that future generations would be interested. Judging by the interest in the firm's surviving military products, they were correct. However, in spite of its comparatively good documentation, there are still many areas of Horstmann's production which are only lightly documented. The scarcity of original source records in these areas raises the importance of surviving firm products such as swords, labeled hat boxes, and a variety of other materials, which assume an important place in this study.

FATHER & SONS

William H. Horstmann, 1785-1850 The founder of the business.

William J. Horstmann 1819-1872

Sigmund H. Horstmann, 1821-1870

The. Horstmann Brothers

Sigmund Horstmann and family. Cartes de viste. (Author's collection.)

Introduction

July 1948. Duncan Campbell, a teacher at Harrisburg Academy, has obtained leave to attend an auction in Philadelphia. The auction takes place in the now vacant and cavernous building of the Horstmann Company at 5th and Cherry Streets in Philadelphia. As the auctioneer proceeds from room to room, he sells the merchandise in each location to the highest bidder. Because this is a bankruptcy sale, creditors will have to agree to the price realized before the merchandise can actually be picked up a day or so after the sale. Hence Mr. Campbell, who must return to Harrisburg, cannot bid on the three stacks of Horstmann firm records piled in a corner of one room, each over six feet high. In them are ledgers, correspondence with Civil War officers to obtain corps badges, copies of the 1857 pictorial uniform regulations printed by Horstmann and many other records which would have made this work a much thicker volume. The papers go to a scrap dealer for $37.50.[1]

The foregoing proceedings represented the funeral ceremonies of the once great Horstmann Uniform Company, the successor to those Horstmann companies which led the military manufacturing trade in the United States during the 19th century. While woefully mismanaged in its last days, the demise of Horstmann represented more than the fate of the single company. Like most of the other great military goods and uniform manufacturing houses, Horstmann had tottered up to World War II and then joined such firms as Lilley-Ames, George Evans, Ridabock, and many others, in bankruptcy. Other smaller firms simply died off or merged with their even smaller successors.

Times had changed. No longer did volunteer militia companies whose members constituted the wealthy elite of communities vie with each other to create a spectacular parade, wearing uniforms they themselves had purchased from private suppliers like Horstmann.

[1] National Archives. Philadelphia Branch. District Court of the United States for the Eastern District of Pennsylvania. Bankruptcy Docket 22693.

The fraternal organizations whose members were also affluent customers of these business houses were in decline. U. S. military officers still purchased their own uniforms and equipage, but most often these were obtained through a standardized PX system, or from individual military tailors who operated around Army posts or Navy yards in a small way. With these changes, large military department stores with their own manufacturing capabilities were no longer profitable.

In late 1815, as William Henry Horstmann arrived in Philadelphia from a Europe just recovering from the Napoleonic Wars, such changes were far in the unknown future. Horstmann had been trained in his family craft as a passementier, a narrow lace weaver whose products included ornaments made out of cloth woven from gold or silver thread. Horstmann was subsequently seasoned by the factories in Paris after a somewhat abortive military career. He was a native of the state of Cassel, because at the time there was no Germany. With the end of the Napoleonic wars he faced an uncertain future in Europe. Horstmann sought to make his mark in a Philadelphia sympathetic to both the weaving trade and to Germans.

Horstmann's ultimate success depended upon his craft knowledge, supplemented by the new factory conditions occasioned by the Industrial Revolution. He arrived in Philadelphia at a time when the economy favored setting up a business based on his thorough technical knowledge of the weaving trade.

The Horstmann firm's significance lies in its ability to transform a craft into a large-scale manufactory and in applying the technological advances of Europe to an American situation, thus creating a large, dynamic and successful firm. As we will see, William H. Horstmann was able to utilize his both European experience and his continuing connections to play an important role in this transition. His craft background was an asset to his ultimate success, rather than a constraint. The equipment, such as Jacquard looms, which he applied to his trade, were claimed to have been the first utilized in America. Horstmann was not unique in having these skills, only better connected and luckier than others.

This book will examine the factors in the success of the Horstmann family business in the 19th century and the decline of the company they created in the 20th. The Horstmann firm would exploit its opportunities to the hilt in its early days, but would ultimately decline as changed times no longer favored its trade or its existence. Through the Horstmann firm, we will view the development of a military goods manufacturing and dealing trade in America, as well as the application of the techniques of the Industrial Revolution to goods formerly produced by individual craftsmen. It is the intent of this work to show the evolution of the military goods trade within the context of 19th century business. As the largest firm in that military trade and the only one for which any appreciable body of records has survived, it is appropriate to deal with the business of importing, manufacturing and retailing military goods through William H. Horstmann and the company he founded.

Columbia Manufactory from printed labels in Horstmann scrapbook.

Table of Contents

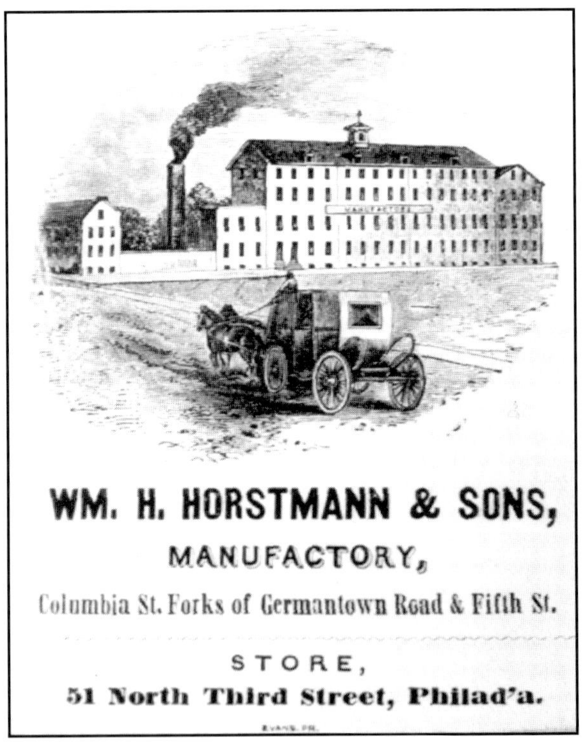

The Horstmann businesses in the mid 1840s
51 N. 3rd Street. August, 1846. Library Company of Philadelphia
51 N. 3rd Street and the Columbia Manufactory from cards printed in May, 1848
Horstmann label books, Historical Society of Pennsylvania

Chapter 1. Young Horstmann

William Henry Horstmann, like his father Henry, his grandfather, and his great-grandfather, was a lace weaver, literally a braid weaver and ornamenter in cloth, or as the trade called it, a "passementier".[2] Craftsmen like the Horstmanns produced the narrow woven textile fabrics used in coach laces, trimmings, and other civilian applications. However, their trade included such military trimmings as sword knots, epaulettes, hat trimmings, as well as other applications of silk, worsted, cotton, and the silver and gold cloth widely appreciated at the time.

He was born in Cassel, the principal city and capital of the state of Hessen-Cassel in what is now Germany on August 16, 1785; the son of Henry and Dorothea Horstmann.[3] The lives of European craftsman fol-

---

[2] Morris, Charles: *Makers of Philadelphia* (Philadelphia, L.R. Hammersley & Co, 1894), p. 66.

[3] This date appears on Horstmann's tombstone as transcribed in 1937. The stone is now partially overturned, but the name and the date 1785 can still be read. St. Michael's Cemetery, Germantown. The family genealogy is listed in Charles Morris, *Makers of Philadelphia*. Confirmation of the family trade and parentage of W.H. Horstmann can be found in Helmut Thiele, *Feinwohner und Familien der Stadt Kassel* Band 6 Geborne H-L published through the Gesselschaft fur Familienkunde in Kuressen und Waldeck. These records give "Johann Heinrich Wilhelm Horstmann" as the son of "Johann Heinrich Horstmann", master passementier born August 25, 1785. This probably reflects the date the birth was entered in the state records. In German usage of the time the first name, in Horstmann's case Johann, was often honorific and not commonly used in everyday address. The same source, while it lists no brothers for W.H. Horstmann, does note the birth of a sister, Sophie Magdalene in 1788. In December 1815, a daughter was born to Johann George Israel Horstmann, another passementier.

lowed an established pattern. William would not simply be allowed to learn his father's trade and continue in his father's business in Cassel. William was apparently not the first son and as we will see, a George Horstmann is known to have done business with the Horstmann firm from Cassel in the 1820s. He may have been an older brother. In any case, William had to complete his trade knowledge outside of Cassel as a journeyman, literally someone who traveled in pursuit of a trade. William then continued his education, possibly in Austria and certainly in France.[4] While it may have been possible for William H. Horstmann to return to Cassel as a master of his trade, he never did so.

The Cassel of Horstmann's youth was anything but tranquil. By Horstmann's twentieth birthday in 1805, Napoleon had conquered much of Europe, including the German states. In 1806 Napoleon installed his brother, Jerome, as king of the new state called Westphalia, formed out of the wreck of the now defunct Holy Roman Empire. This new country included Cassel and with it, presumably the Horstmann family. The new state proved to be the most popular and effective of the French client kingdoms because, for the first time in many years, its citizens were fairly and competently governed, but it was almost constantly at war. In the spring of 1809, an attempt was made to overthrow Jerome Bonaparte by a Major Ferdinand von Schill, a Prussian officer. This attempt failed when Schill died while besieging the fortress city of Stralsund, and his followers were disbanded. Napoleon was usually magnanimous toward defeated enemies, but this time the defeated troops were treated as traitors. Many were hung, and others forced into hard labor.

All this is detail except that William H. Horstmann is cited in his firm's history as having been a part of the Schill expedition.[5] It is difficult to reconcile Napoleon's actions toward Schill's army with what the firm's commercial history claims as Horstmann's next move. He is cited as the foreman of an unknown French factory specializing in the textile weaving trade, and a woven portrait of Napoleon is pictured in the book

---

This may be the same individual who manufactured a gold curl and box for W.H. Horstmann in the 1820s (see below). However, no record of this individual's birth was noted in Thiele. From the date of his daughter's birth, George Horstmann was probably a contemporary of William H. Horstmann. If not a brother, George Horstmann was likely a close relative.

[4] *One Hundred Years 1816 - 1916 the Chronicles of an Old Business House in the City of Philadelphia*, William H. Horstmann, 1916, pp. 8-11.

[5] *Ibid.*, p. 10.

as having been made by Horstmann in those years.[6] Whether or not this history is completely accurate, it does reflect the changing nature of European society at the time. Horstmann, as a German apprentice or journeyman was able, in the freer social system of Napoleon's empire, to acquire a good technical background and possibly to have risen to a position of some responsibility. This explanation gives a plausible source for Horstmann's knowledge of how to set up a textile factory with the latest equipment, which he ably demonstrated in Philadelphia in the middle 1820s.

Horstmann probably remained in France from approximately 1810 to 1815. One of Horstmann's R. G. Dun and Company ratings states that he was a veteran of Waterloo. This later service is possible, but is cited nowhere else, and is probably only a garbled reflection of Horstmann's service in the Napoleonic wars.[7]

From all this, certain suppositions seem plausible. Somewhere around his twentieth year, about 1805, Horstmann left Cassel to continue his trade education. During the next ten years, he learned his trade well, including the use of new and dynamic manufacturing equipment such as braiding machines and Jacquard looms. He seems to have been caught up in the Napoleonic wars as a soldier on at least one occasion, and possibly on several. Horstmann's career in Paris, while only very sparsely documented, would have been a logical place for a journeyman ex-soldier to have gone to learn and practice his trade, as France's war industries were in high gear at the time and Paris was one of the centers of the French textile trade. Horstmann's decision to relocate to Philadelphia a few years later was a correct career decision, as well, since Philadelphia was also developing a textile industry and was friendly to German immigrants. In later years the Horstmann Company would maintain a purchasing branch in Paris, and a manufacturing plant in Lyon.[8]

The economic conditions in Europe in the early 19th century may help to explain Horstmann's travels through Europe and his ultimate

---

[6] Connelly, Owen, *Napoleon's Satellite Kingdoms*, Free Press, New York, 1965, p. 204, and *ibid.*, p. 13.

[7] R. G. Dun and Company Collection, Philadelphia 1.511, entry October 6, 1861 "The old man was in the Prussian Army at the Battle of Waterloo."

[8] *Philadelphia Yearbook 1917*. Philadelphia Chamber of Commerce, 1917, p. A47. The manufacturing plant is probably a late addition to the Horstmann capabilities. However, a regular Paris office was a feature from at least the mid-19th century.

decision to emigrate to America.[9] The Industrial Revolution brought many changes to Europe. No longer did the craftsman who actually produced merchandise respond personally to the needs of an individual customer. Instead, goods were now made in impersonal factories rather than in individual workshops. Merchandise was shipped to markets which could often be a distance away. These industrial changes came in slightly different forms to each European country. Unlike England of the early 19th century, the craft guilds were still strong in Germany at that time. This situation may have been the result of Germany's patchwork of free cities and divergent principalities in what was then called the Holy Roman Empire, a remnant of Germany's Medieval governance. Equally important, Germany remained even more overwhelmingly agricultural than France and England. Both factors emphasized Germany's somewhat arrested political and industrial development in the early 19th century.

In France at an early date, certain factory enterprises had been state sponsored. These usually consisted of the manufacture of luxury items such as ceramics, ornamental textiles and arms. The manufacture of such items also appealed to the rulers of various German states. Large manufactories of porcelain were set up in cities such as Dresden, while the arms industry of Solingen was carefully protected. But these were exceptions. In Germany the pace of industrialization was much slower than in England or France.

Throughout Europe, at whatever pace, the Medieval idea of the craft guild was slowly giving way to the concept of the capitalistic factory enterprise. While there were free masters who joined no guild, the guild organizations tended to represent the craftsmen in cities, not the rural economy. Additionally, the amount of specie available worldwide had been growing markedly from the 15th century onwards through the silver and gold discoveries in the New World and in Austria's Hartz mountains. There was now far more venture capital to fund new enterprises than there had been at earlier dates. Transportation was also much improved.

All this tended to foster a climate of economic expansion and a search for new markets. The prevailing economic theory, mercantilism, saw control of these expanded markets as a source of patriotic protection and security for the nations who monopolized them. The idea of mer-

[9] See *The Cambridge Economic History of Europe* (Cambridge: Cambridge Union Press, 1977). 8 volumes. Vol V pp. 44, 466. This history gives a good synopsis of the economic conditions in Europe during the Napoleonic era.

cantilism was evolving at the same time as that of the modern business. A "firm" in this new usage functioned as a legal entity, an accounting entity, and a credit entity rather than through a craft production system. All through Europe, enterprising craftsmen were "putting out" raw materials, often in rural areas, and getting back semi-finished products.[10] The craftsmen who received these half-finished items and completed their manufacture process often became merchants or factory owners. In Solingen, for instance, the individuals involved in the sword making process were the sword smith, temperer, and sword furbisher (reider). The latter actually put the swords together, finished them, and thus ultimately marketed the swords.[11] Merchants, not craftsmen, could reach out to foreign markets.[12] What was required for such ventures was capital and an ability to ship goods in large quantity to far away markets, something the guild structure had not contemplated. However, as this capitalist process continued, a few guilds in larger cities took on these mercantile techniques as way of competing with the new system, but they were unusual.

In addition, new ways were found to expand credit. One of these was through the "mercantile draft", an 18th century concept which continued well into the 19th century. Debts were paid with notes rather than cash and these might circulate for months, collecting as many as a hundred endorsements in their travels and serving as a sort of alternate currency. These notes were usually discounted from 10-20% depending on the perceived ability of the original writer to have them redeemed quickly.

The new system was hard on individual craftsmen. In a factory system, masters might become foremen in a factory and journeymen became skilled workers, while apprentices were relegated to the status of factory hands. Just as markets became impersonal, the family concept of master and apprentice gave way to a modern employment methodology. As crafts competed for far away markets, guilds tended to break down. In Germany a bewildering variety of customs barriers imposed by the small principalities tended to put a damper on the ability of individual enterprise to approach foreign markets. Great craft centers like the sword blade manufacture in Solingen and the ceramics of Meissen

---

[10] *Ibid.*, p. 469.

[11] *Ibid.*, p. 470.

[12] Cameron, Rando: *Essays in French Economic History* (American Economic Assn., 1970). "French Gilds under the Old Regime", p. 254.

as well as textile manufacture in certain defined areas developed world wide markets through the excellence of their goods. However, Germany as an entity could not compete with the quantity and aggressive worldwide marketing of English and French goods produced in the early 19th century. In early 19th century Germany, many finished products continued to be marketed at craft fairs, reminiscent of Medieval guild fairs, rather than by consignment trade as English and French merchants were regularly doing. Thus the guild system remained relatively strong in Germany after it had weakened in more industrialized countries like France and England. This conservative attitude towards the guild structure in Germany may account for Horstmann's having to continue his education outside of Cassel, commencing a series of journeys which led ultimately to Philadelphia.

The Napoleonic union of Germany pointed the way to breaking down the barriers imposed by the patchwork of German states, but after the fall of Napoleon and the restoration of the old Germany, it took years of customs unions, confederations, and ultimately the inception of a new German state to completely dissolve them. Industrialization in Germany at the end of this process in the later 19th century took on an authoritarian form of state sponsorship working along with private venture, a system different than that utilized in England and the United States.

Textiles in particular were easily produced in factories. Some of the earlier 17th century factories in England and France were engaged in the production of textiles. In Germany, textiles also represented a staple manufacture, particularly in the Middle Rhine to the Meuse. Power looms were in use in Danzig in the early 17th century but were ultimately prohibited through influence of the guilds. The Dutch and Swiss took up this manufacture and one 17th century Dutch loom was capable of weaving 24 ribbons at once. Such machines were in use in Germany as well. Indeed, immigrants from the Netherlands stimulated the "passementier" (braid weaving) trade in a number of German cities. This trade was the traditional craft of the Horstmann family and of William H. Horstmann himself. As early as 1605 Cologne had a guild of passementiers which numbered 2600 masters.[13]

The major innovation in the textile trade was the introduction of the so-called "Jacquard loom" about the year 1800. Joseph Marie Jacquard was a weaver in Lyons, France, and invented, or more properly refined, an already existing system by Vaucanson. In its final evolution,

---

[13] *Op. cit.*, pp. 470-471, 521.

Model of hand loom made by William J. Horstmann. One Hundred Years, *p. 17.*

the Jacquard loom resembled a primitive computer activated by a series of punch cards which controlled the threads woven by a power loom. These cards were in sequence and their use produced an automatically repeating varied pattern in the finished textile.[14]

In spite of the changing nature of European economics, America must have looked like a more fertile field for the employment of W. H. Horstmann's talents. At the very least, the economic readjustments at the end of the Napoleonic Wars and the need to re-establish himself among many ex-soldiers now eager to become tradesmen must have influenced Horstmann in his decision to leave for America. The conservative Europe created as a result of Napoleon's defeat would now proceed to reestablish the fragmented Germany of Horstmann's youth, bringing back the unpopular Elector of Hesse-Cassel, a prospect which must have made returning home less than congenial. Even if he did return to Cassel, the family business was probably being run by other members of the Horstmann family. Accordingly, William H. Horstmann still would have had to fight for a place in Cassel's textile weaving community, whose craft guild most probably would have been less than interested in Horstmann's newly acquired skills at running a factory.

---

[14] Figuirer, Louis: *Les Grandes Inventions Anciennes et Modernes* (Paris: L Hachettend Co., 1867), pp. 407-12. See: Chapter 3 below, pp. 76-78, for an explanation of the specific machinery used by Horstmann.

Towards the end of 1815, or early the next year, Horstmann sailed for Philadelphia. Like his stay in Paris, his actual passage is largely undocumented. No passport, ship manifest, or other document listing his passage has so far been located. The actual date of Horstmann's arrival is taken from family business tradition and such events as his Declaration of Intent to become a U. S. citizen, his city directory listing, and the birth of his daughter in 1817. The matter was apparently confusing even to William H. Horstmann's grandchildren who variously gave the date of the firm's establishment as 1815 or 1816. As we will see, Philadelphia with its quickly growing textile trade and large German speaking population would be a most congenial spot for Horstmann's own relocation. He was no longer young in 1815. He was 30, then unmarried, presumably of some means, and without obligations.

In America, Horstmann was careful to give his two sons a knowledge of German and an interest in German affairs which would last through their entire lives. In this process W. H. Horstmann returned to Germany several times on business, once with his eldest son. Given his continued trade contacts with Germany and his pride in showing his elder son the places of his youth during their later business travels together, there is every reason to suppose that Horstmann remained interested in Germany and in contact with his family. At the same time the Germany of Horstmann's mature years had regressed back to a patchwork of small states each with its own restrictions on travel and customs barriers. Because this political situation was inequitable, it necessitated a series of customs unions in the second quarter of the 19th century, then sparked the rebellions of 1830 and 1848, and finally led to German unification in 1871. William H. Horstmann would only live to see the initial steps in the process of German unification, but with his commercial and social contacts continuing throughout his life, the conditions in Cassel were probably never entirely out of his frame of reference.

Chapter 2. The Military Goods Trade

The United States and Philadelphia were a logical destination for a man of William Horstmann's talents. More than a century before Horstmann's arrival in Pennsylvania, William Penn had envisioned his colony as a haven for the oppressed, and accordingly had invited both English and non-English speaking groups to settle within his Pennsylvania. Most prominent of these non-English speaking peoples were the victims of Louis XIV's territorial ambitions of the late 17th century in the Rhineland. These struggles, contemporary to Penn's establishment of a haven for religious diversity in America, had produced untold misery for the closest of France's neighbors to the north and east. Many of Rhenish Germans accepted Penn's invitation to come to Pennsylvania while other German refugees settled in rural New York.

While today's Pennsylvania German population is best known in Pennsylvania's rural areas, in fact many of the immigrant Germans were artisans or merchants who stayed in larger cities, especially Philadelphia. Among the trades they brought with them was that of weaving. Through the 18th and 19th centuries Pennsylvanians of German descent remained well established in most of Philadelphia's trades and manufactures, though perhaps less so in shipping and mercantile enterprise. Many prominent Philadelphia names, among them Rittenhouse, Wistar, and Bringhurst, were the product of the amalgamation of German culture and English commerce during the 18th century.

The 17th and 18th centuries were the era of the merchant. Companies of gentlemen adventurers and soldiers explored or conquered large portions of Africa, India, North and South America, and China. Many of these explorer entrepreneurs were in fact commercial firms which, like their smaller cousins, answered to investors who expected a return on their money. Often the capitalist got none, and the colony business remained a risky enterprise. To the European mind social status still involved the ownership of land rather than concentrating capital solely upon the growth of business as would motivate its 19th century successors. A successful 18th century merchant might hope to be a country squire at home or a plantation owner in the colonies. The artisan craftsman could have no such aspirations because he would never accumulate sufficient wealth to purchase an impressive quantity of land, if indeed any land at all.

The economic policies which governed most nations were "mercantile," a term which referred to owning colonies in order to enrich the mother country. Industry was systematically discouraged in all colonies, as was the importation of hard coin which could serve as capital. In one respect the British North American colonies were different. American entrepreneurs desired to set up industries if they were able to do so. One industry was relatively common in colonial America: ironmaking. While early iron works like those at Saugus, Massachusetts, and the later innumerable iron furnaces in Pennsylvania, were frowned on by London, they nonetheless survived. Other industries such as cloth making and brass goods only got their real start in the years after the Revolution.

Most manufacturing in both Europe and America before 1790 was conducted through trades such as gun smithing, printing, cabinetmaking, etc. These trades in Europe were traditionally controlled by guilds or regulated through government monopolies. While the guilds did not enjoy similar power in Federalist America and government monopoly was not a factor at all, the horizon of the early 19th century American artisan was not without its own special perils. Foreign imports kept profits low. A growing fear of industrialization was perceived as a threat to the craftsman's freedom and to his ability to make a living. By the second quarter of the 19th century this threat would become real.

The primacy of the merchant as the leading colonial citizen did not change with the Revolution. In spite of the politically wise efforts of Alexander Hamilton, many merchants still regarded themselves as a better class than the tradesmen, an attitude which helped to bring about the downfall of the Federalist party.

In this era it was the merchants, not the artisans, who got rich and lived in large houses. Tradesmen were paid little for their labor. Much of the value of their product was in the materials used and in their cost of transport.

Merchants, on the other hand, had their own ups and downs. Trade was a definite risk and involved taking chances. The state of international relations in the era of Napoleon often led to situations where one country attempted to disrupt another's trade. In America, mercantile problems came to a head in 1807 when an embargo imposed by the United States on foreign goods was replaced by a "nonintercourse act," which helped no one except Napoleon, who used it to his political advantage in pushing the United States closer to war with England. The disruption of the normal import trade led to a certain prosperity in American produced goods which, without foreign competition, were now in demand. Domestic manufacture was in turn jeopardized when peace returned in 1815. That peace was accompanied by the usual hard times, compounded by an attempt on the part of the British to recapture the American market by dumping all sorts of cheap goods on our shores.

Horstmann's competitor, Henry Korn, a fellow immigrant, has left us a long statement of his woes in the manufacturers' census of 1820:

> "As respect the actual and past condition, I would beg to state that in the years 1814 and 1816 the sales was about three times what they are now, for since the general peace in Europe, the manufacturers of Tapestries, Military articles and &c who were then employed by the European armies, losing at once their resources, on finding a market open to them in the United States, literally glutted it so that upholstery trimmings have been for some time sold lower at our . . . establishments and by private sale than they could be manufactured for either in England, France, or Germany, to the great detriment and almost inestimable ruin of the same class of manufacturers in this country, in the year 1814 I had three men and four boys and nine women in my employ, all with cheering countenances, fired with the stimulus of business, but now it is quite the opposite, for even what little is doing is reluctantly done because it is lackly wanted, as respects coach trimmings that is very little in demand now owing to the general depressed state as almost all who formerly kept carriages or gigs now have them to dispose of. Till the year 1817 every country merchant that came to Philadelphia to purchase goods wanted tapestries but now they are forgotten. It wil be remembered that Wool, Cotton, and Flax are staple articles of this country, and we can also manufacture from these same materials all kinds of military trimmings necessary to be

used in the United States, but for that those same articles would have cost enormous prices during the last war with Great Britain and would have been in all substantially manufactured by the very enemies with whom we were contending. . . ."[15]

Henry Korn's difficulties with foreign manufacturers reflect the context of Horstmann's difficulties in starting a new business in a depressed economy, having to fight both foreign monopoly and domestic difficulties.

Merchants would import what sold well. Often goods were advertised for sale as fast as they could be imported. Imports competed both with fledgling American industry and with the trades of individual artisans. The gun trade of the later 18th century initially came off comparatively lightly in this struggle. Many gunsmiths were well established in their communities. Because some of the materials needed to fabricate guns were available thanks to the iron industry, gunsmiths could, and did, produce a respectable product in the early years of the trade when it was necessary to fabricate entire arms by themselves. However, by the early 19th century a flood of commercial gunlocks and the establishment of forges which specialized in making barrels assisted the local gunsmith in establishing a cheaper way to produce his finished product by furnishing already made parts for him to include in finished weapons.[16] While finished parts helped the gunsmith produce his own wares, the fully assembled English and Belgian arms competed aggressively with our own products during the first half of the 19th century.

The industrial revolution also strained and ultimately changed traditional society. In England, which was far more industrialized in 1820 than the United States, many workers blamed the post-Napoleonic recession on the machines themselves. This culminated in the so-called Peterloo massacre, where the workers attempted to smash machines which they saw as dehumanizing and starving them.

As Henry Korn noted, in America fears of the social effects of industrialization were for the moment secondary to the fear of cheap imports. Calls for a protectionist tariff had gone out from both tradesmen and manufacturers soon after the Revolution but did not meet an overwhelming response in Congress. Then as now, consumers wanted to be free to

---

[15] *1820 Manufacturers Census of the United States.* Pennsylvania. Philadelphia County, no. 587.

[16] Brown, M. L.: *Firearms in Colonial America* (D.C.: Smithsonian Press, 1980), pp. 244, 272.

purchase the cheapest goods they could, an argument which appears to be timeless.

The problems of cheap competition from Europe were by no means restricted to the gun trade. Items such as swords, uniforms, insignia, headgear, etc., also were routinely imported, along with huge quantities of English ceramics and cloth which for a time dominated the American market. After all, England had won the trade wars with France. Why should she not get rich?

Perhaps typical of the problems faced by the American manufacturer in the face of foreign imports is the plea of Nathan Starr cited from the Census of Manufacturers in 1820. In this census Starr claimed that he had $50,000 tied up in capitalizing his business, paid $7,500 in annual wages, and had $5,500 contingent expenses in order to sell swords for $4 and $5 each. It was a time of widespread depression and all manufacturers struggled to sell their goods. Starr states:

> "During the war employed 84 men. had a ready sale for all the swords they could make they now employ but 15 men on contract for the gen'l government and have no sale for surplusage in consequence of English & French swords of inferior quality and the refuse of their workshops flood over this country at any price that can be named for them.
>
> "This establishment had commenced making fine swords but have been obliged to abandon it and have made none of late except in a few instances where they have been ordered by the general or state governments."[17]

In actuality many of Starr's problems were of his own making. His product was essentially a utilitarian enlisted man's weapon although Starr attempted more grandiose officer quality and presentation pieces. His shop could not compete with the quality of foreign arms which included many varieties of swords for both the Federal officer and the volunteers.

Be that as it may, military goods and arms were obtained in a variety of ways. Here are a few practical examples of how the soldier got his arms and accouterments. The first method was by directly purchasing the items from the manufacturer.

---

[17] *Manufacturers Census of the United States*. 1820. Connecticut. Middletown. Entry 213, Nathan Starr.

An invoice of Le Page, a French arms dealer, made out to Lt. George Boyd in 1814 is in the Daniel Parker papers in the Historical Society of Pennsylvania. In it Boyd is purchasing directly in France a gladiator model sword, morocco leather belt, crimson sword knot, and Colonel's epaulets. The quality of these pieces sold to a Lieutenant is perhaps the reason that the bill turns up in the papers of Parker, the Adjutant General of the Army. It is just as likely the pieces were purchased for Parker by Boyd.[18] Such direct purchase was unusual, though state governments would sometimes turn to foreign arms producers in the years after the Revolution.

The failure of Pennsylvania's arms contract with the English firm of Ketland pointed up the dangers of relying on foreign firms to provide vital arms in wartime. The English government simply stopped the shipment of these arms because of their own difficulties in the early 1790s with the successful French Revolution. The Federal Government decided to set up its own armories at Springfield and Harpers Ferry and to supplement their production through a system of domestic contracts.

Individual officers such as Consul William Jarvis, who needed a U. S. Navy uniform in January, 1803, to attend the christening of the heir to the Portuguese throne (diplomats not having their own until 1816), sometimes took direct action. According to Jarvis' daughter, he simply ordered his directly from London.[19] The uniform survives and indeed its buttons are part of the US 1798 series known to have been produced for Carter/Lombard St. in London.

The John J. Low firm advertised in the Boston *Atlas* of 12/11/1840. The ad shows them obtaining from "Manufacturies at Paris & Lyon, Lyon Gold & Silver laces, epaulettes , stars and straps. Paris made silk sashes, worsted do, swords, chapeau, belts, chains, buttons, Navy swords, epaulettes."[20] In the *Columbian Centenel* of May 4, 1825 Low has received

> "From Liverpool, 8 packages, containing swords of almost every description, holster and pocket pistols; belt plates, cap do; spurs; plume sockets, sword hooks; rings; sword and cap chains; ball buttons; stamped do of various kinds &c &c. From Havre. 6 packages containing Epaulets; wings; laces; tassels; braids; stars;

[18] Daniel Parker Papers. Manuscripts Library. Historical Society of Pennsylvania.

[19] Cutts, Mary Pepperell Sparhawk: *The Life and Times of Hon. William Jarvis of Weathersfield, Vermont* (New York: Hurd and Houghton, 1869), pp. 160-161.

[20] *The Atlas* (Boston), November 11, 1840.

rosettes; Vulture and common plumes of all colors and lengths; with a great variety of other fancy and ornamental trimmings in the military line . . . ."

The ad also includes candlesticks, jewelry, watches, plated ware, and jewelry: "All of the above are fresh and new and were imported direct from the manufacturers or agents . . . ."[21]

Only one piece has been located from Low's extensive import trade, even though the firm still exists as Shreve, Crump & Low, a Boston jewelry firm. A bellcrown cap went to a member of the Bolton, Mass. Light Infantry in 1831.[22] While the insignia could be imported, it is unlikely that the cap body was made very far from Boston. Merchants built their trade around what they could sell, not whether they were importers or simply conveyed local goods to a customer.

Another message which seems to be evident from the Low advertisement is that during the first quarter of the 19th century, metal goods were imported from England and woven items from France. There appears to be a good deal of truth in this statement, given the known national origins of the demonstrably immigrant tradesmen who set up business in America. In fact, this process is not quite so simple when we remember that at the same time, England was getting most of its own sword blades from Solingen, which had established a world trade in sword blades. Solingen blades were still used in British military weapons as late as the 1860s and 1870s.

Before we go too deeply into the transition from merchant to manufacturer in American arms, we do need to look at the customers who after all chose which items to purchase–the militia and regular Army which made up the military establishment in the United States.

A painting of the New England Guards encampment done in the 1830s hangs in the Bostonian Society. In it the militia saw themselves as neat and orderly, true soldiers. In fact, the painting is of an encampment of a volunteer militia organization. The volunteer militia was made up of company sized units consisting of persons who had enlisted to drill and undertake military duties on a regular basis.

Hundreds of volunteer companies with colorful names such as the Troy Citizens Corps, the Boston Light Infantry, Savannah Volunteer

---

[21] *Columbian Centenel*, May 4, 1825, p. 4, col. 3.

[22] The cap is accompanied by an 1830 bill to Ensign Joel Sawyer of the Bolton Light Infantry. Inside the Sawyer cap is a label by the Low firm. The piece was in the Bolton Massachusetts Historical Society but has been stolen.

Guard, Richmond Light Infantry Blues, and a number of units called the Washington Greys were the direct ancestors of both our present national guard and our state police. However, there was a dark side–the "common militia", a relic of the Anglo-Saxon "fyrd", an armed band of free men owing service to their lord, in this case their state. This force, literally all of the free, physically able men in the state, met on training days–a few a year, mandated by law. This militia service was compulsory. While the government armed militia units after 1808, a militia company required no enlisted uniforms until it was actually taken into Federal or state service in an emergency, and it got none. The fame of the common militia rested on Bunker Hill, Saratoga, and Kings Mountain in the Revolution, though perhaps not on disasters such as its rout at Camden, S. C. It heaped infamy upon itself by non-action during the war of 1812 and thereafter quickly became irrelevant. Its demise was satirized by cartoons such as "The Militia Muster" and the "Nations Bulwark" during the 1820s, which showed stereotyped characters such as the strutting banty-sized officer in full regalia and inept soldiers, usually both drunk and uninterested in the proceedings.

The post-Revolutionary regular army was both small and popularly distrusted. The common militia was both unwieldy in size and untrained, hence the smaller and far more efficient volunteer militia became the nation's ready reserve. Within loose guidelines, these volunteer units dressed as they pleased. They were the most important customers of the military goods dealers, manufacturers and importers. While officers of the "common militia" did have uniforms, these were most often U. S. Army patterns. Militia troops were armed by the United States government, but were not uniformed.

By the Militia Act of 1792 which formalized the whole militia structure, militiamen were to attend training armed with their own weapons. This system did not work because many men did not own arms. The problem was addressed by individual states through private purchases such as Pennsylvania's 1798 musket contracts with gunsmiths like Evans, Miles, Henry, and others. A few states retained arsenals where they manufactured their own arms. The situation was partially redressed in 1808 when the U. S. government agreed to supply the militia with Federal arms, cartridge boxes, and ammunition. The state contracts then either ceased or were relegated to special purchases, such as rifles procured during the War of 1812.

The Federal contracting system prior to the Civil War attempted to meet as much of the need as possible for arms and accouterments for both the Federal Army and the militia through arsenal production. Contracts were often let for additional arms and equipage.

The uniform fashions and patterns of arms copied by the United States were European in design, and as we have seen, sometimes European in manufacture as well. The French were still regarded as the dominant land power in Europe in spite of their habit of actually losing their wars. Accordingly, they were held as arbiters for patterns of both swords and uniforms by many countries. U. S. muskets adopted after 1792 were direct copies of the popular Charleville to which our troops were introduced during the Revolution. The same was true for both swords and military uniforms through 1870. Certainly the French turned out a good product which was held in high esteem regardless of actual French performance on the battlefield.

Thus, European, and especially French, fashion ruled supreme in the minds of American soldiers. The patterns employed in making arms, the cut of the soldier's coat, the design of the insignia on the soldiers' hat were influenced by European styles. Contemporary fashion in turn employed classical art forms or looked to outlandish soldiers who came into prominence during the Napoleonic Wars or later colonial conflicts, such as the Hussars and the Zouaves. The same forces had an influence on decorative arts and furniture design as well. The eastern European czapka shown in Margerand's great work on French headgear was adopted by Napoleon's guard from the Polish lancers.[23] It turns up on the heads of the Independent Corps of Cadets in Boston between 1816 and 1840. French design lost popularity in the United States after their defeat in 1870. By 1872 our soldiers were wearing "Swedish" spiked helmets and double-breasted coats.

The Roman imperial eagle used by several English units comes from the Napoleonic standard captured at Barossa, Spain by the 87th Regt. in 1811. It flew over the seas to be incorporated into our dragoon cap insignia in 1833.

Perhaps the most elegant uniform of all was that of the Boston Hussars of 1810. This company was made up primarily of rich merchants and other Boston gentlemen. It is reported that the horse and uniform cost $1,000–no mean sum in 1810, but they were certainly gorgeous. The Hungarian irregulars who initially introduced the Hussar into European consciousness could only dream of such splendor. Some changes were,

---

[23] Margerand, J.: "Les Coiffures de l'Armée Française". *Revue Mensuelle Illustrée* (1909-1924), vol. 2, p. 63.

however, practical, such as the false crown introduced by Wellington's troops to give the illusion of height without the weight. On the other hand we have the bellcrown cap. Born in Russia, it somewhat resembles a black mushroom. It was adopted by the U. S. Army in 1821 and lasted until 1832, except within the U. S. Marine Corps, which retained it a bit longer. At about this time the Army came up with fatigue caps for undress wear because no one was going to march long distances in a bellcrown.

By the second quarter of the 19th century society itself was turning from a craft-based system of individual manufactures into an industrial, mass production-oriented system. The individualistic craft mentality had encouraged freedom of design and construction. The factory fostered conformity and standardization of output. In the early years of the republic, the production of weapons, insignia, uniforms and other U. S. Army military goods was governed by loose specifications. These allowed for the variation which society in general accorded its craft produced products. This tolerance tightened up gradually from the 1840s onward, until by the Civil War manufacturing was capable of the standardization which we now expect from all products and which our forebears did not similarly expect.

We can now revisit the import merchant and the manufacturer. We have seen the merchant as largely successful and prosperous through the first decades of the 19th century and the manufacturer as relatively small, reflecting a somewhat dubious official approval during the Colonial period. Much of the actual productivity was in the hands of artisans whose labor counted for little as compared to the cost of the materials used in making the product. The real money went to the merchant importer in the 18th century, as it was to go to the industrialist in the later 19th. The first half of the 19th century was in fact the transition point between the primacy of the importer and that of the industrialist.

In the 1790s the emphasis on imports began to change slowly. Successful cotton mills were set up in Rhode Island, and the first rolling mill in the United States opened its doors in 1792 near Boston. It was owned by Paul Revere, ever a better entrepreneur than a soldier or a horseman.

As an example of what the early manufacturer faced, we have only to look at brass manufacture. The colonial copper mines in Connecticut had played out by the Revolution and they served only as jails to keep Tories locked up. Copper came from used boilers and ship sheathing, hence from Europe. Brass is made by adding calamine (zinc) to copper, the more zinc the whiter and harder the brass. Calamine in the early 19th century was imported from Germany. This situation gradually improved. By mid-century, zinc had been found in Georgia and the great copper deposits which still supply the country's needs had been located in Michigan. However, back in 1804, George Armitage, who made many of the early insignia and button patterns for the Army, complained in a letter to the Commissary of Purchases about a lack of copper fit to roll– in Philadelphia.[24]

The picture was changing in another way. After the Revolution a new generation of immigrants with new and useful skills came to America. They added greatly to the abilities of the artisans already here. Gunsmiths already had imported their trade from Germany and England during the course of the 18th century and were a vital part of the American scene by the Revolution. While they could make some of the arms they needed from locally available iron and wood, new or used manufactured parts were normally imported. Particularly troublesome was a lack of high grade steel, especially the spring steel used to make the interior lock mechanism. The craftsman's problems with brass have already been noted.

Swords were another matter. Making scabbards which would hold, wrapping or checkering the grips, and etching blades were skills which were either scarce or non-existent in America. The beautiful Philadelphia solid brass grips on swords, while following a European precedent, were not worth much if you wanted a functional weapon. However, they were easy to make if you had brass, which was available only by reusing scrap. The rest still had to be learned. While sword makers like the Rose family of Philadelphia were quite competent blade smiths, the rest of their early production might as well have been executed by the Lavender Hill Mob. Early 19th century U. S. contract scabbards are much scarcer than the swords for good reason–they fell apart. Leather wrappings on these productions have quite often come loose from the wooden grips which underlay them. We simply could not make the wire-wrapped grips which were long an important part of European sword making technology.

And so it went. Buttons popped off uniforms because we did not know how to make good shanks. We knew how to plate metals by mixing powdered gold or silver with mercury which volatized in heat, leaving a beautiful finish and a workman dying from mercury poisoning.

[24] Armitage to Tench Cox. May 13, 1805. US Army Quartermaster General Consolidated Correspondence File. RG92. USNA.

No one knew how to burnish the plating–a final step to make it look lustrous. Militia units like the Cleveland Greys had to send to New York for their first uniforms in 1816 because no grey cloth was to be had locally.[25] However, where there was a need, men with proper skills were filling that need. Some of them were joining with capitalists to set up American industries.

We first meet George Armitage in the Philadelphia directories of 1795. He and a brother had evidently emigrated from Birmingham, England. We next find him in *Claypoole's American Daily Advertiser*, 6/11/1799, making uniform eagles for the volunteer companies

> ". . . in silver, gilt or plated, in the manufactory of George Armitage, in Race Street . . . or of J Brearly no 75 north second street; likewise the eagle and anchor for the navy, in gold or gilt, with a variety of fancy eagles and ornaments. . . . N.B. The eagle is marked GA in order that it may be known in the future; every true American ought to recollect it is adopted by General Washington."[26]

It used to be thought that Armitage designed much of the insignia used by the U. S. military between 1800 and 1825, but it is now strongly suspected that Armitage was a player in a larger process. Moritz Furst, an Austrian who emigrated to the United States thinking he would take over the U. S. Mint, designed insignia in addition to executing many of our War of 1812 presentation medals. However, between Armitage, Furst, and the Commissary General of Purchases, Callender Irvine, many of the standard devices in our national iconography were created.

Until his death in 1825, Armitage was the major contractor for U. S. insignia, but he was hardly the only one. William Crumpton, who first settled across the Delaware in Burlington, New Jersey, also had insignia contracts, as did ultimately Robert Martin before the latter went bankrupt in 1805. Martin's manufacture apparently helped to stop the need for officers to obtain their buttons in London. Martin may himself have been an English immigrant and Crumpton was almost certainly from Sheffield. In Waterbury, Scovill, a Yankee, learned the business of fixing a rigid shank on buttons and the secrets of burnishing from Samuel

---

[25] Tibbits, George W.: *A Brief Sketch of the Cleveland Greys Outlining A Few of the Principal Happenings During A Period of Sixty-six Years*. Cleveland Greys (1903), p. 8.

[26] *Claypoole's American Daily Advertiser* (Philadelphia), 6/11/1799, p. 3, col. 4.

Croft, a Birmingham native. In Attleboro, Massachusetts, another center of button making, the trade got its start from Edward Price, an Englishman and from a mysterious figure known as "the Frenchman".

Bartholomew Delapierre, in business 1818-65, was a Dutch national who saw 1812 service as an aide to Gen F. B. Porter, and apparently was wounded in that conflict. He set up a coach and military lace business in New York after the War. It is significant to note that the artisans dealing in the metal trades seem to come from England, and those in the woven ornament industries from the Continent.[27]

Through the 19th century, Philadelphia was a center of the American sword trade. We do not know where the Rose family learned how to make swords. They were Chester County farmers but, like the brass makers of Waterbury, they learned what would make them money, with or without foreign help. Their specialty was making blades only, leaving the hilting to others.

The Roses found a blade etcher named John Meer and they were soon turning out a product good enough to be mounted in many important 1812 era presentation swords. However, the creation of a presentation sword was a cooperative enterprise. Rose supplied only the blades for others to decorate and mount. Rose's efforts to hilt and scabbard their plainer U. S. contract arms were clearly crude because the skills associated with hilt and scabbard manufacture were not within their trade specialty in making high grade blades.

Because the presentation broadside exists for the sword given by Gen. Gaines to Jonathan Goddard Watmough for service at Fort Erie in 1814, we learn a good deal about the early 19th century manufacture of presentation swords which is not readily apparent from the artifacts themselves.[28]

First, because the Watmough sword was not an official "thanks" by the government, but was privately presented, it was actually given in 1830 when there was an occasion requiring its manufacture, namely Watmough's successful campaign for Congress. This was not unusual for the times. Even the authorized Congressional and state presentions of swords and medals sometimes reached the recipient years after the events they commemorated.

---

[27] R. Graham Dun papers. Baker Library, Harvard University School of Business Administration. New York vol. 375, p. 200, A/30, Bartholomew Delapierre. Entry: April 22, 1859.

[28] Broadside describing presentation of the sword by Gen. Gaines to Jonathan Goddard Watmough in 1830. Watmough papers, Historical Society of Pennsylvania.

In producing the Watmough sword, John Hub made and marked the silver hilt. What is not apparent is that Harvey Lewis designed the sword. Lewis was a famous Philadelphia silversmith. Just as the creation of our national iconography was a process shared by a number of hands, we cannot credit Rose with the design and manufacture of great presentation swords. Rose made the blades, competently. Meer decorated those blades while a silversmith, J. Hub, made the hilt. A designer, Harvey Lewis, whose name is usually not on the pieces anywhere, did the conceptual work. Apparently this was often the case. However, when we come to the great jewelry design houses of Fletcher & Gardiner and, later, F. Bailey & Co. and Ball, Black & Co., we can have some confidence that the name on the sword represents the designer as well as the manufacturer.

The Rose family was active primarily from the 1790s through 1820. Their successors changed the basic competencies of the trade as practiced in Philadelphia. F. W. Widmann arrived from Germany in 1816 carrying only his tools. He could make rather showy specimens, but imported his blades from Germany. Unlike Rose, whose competencies were in the manufacture of blades, Widmann and his contemporaries never made a blade, but usually hilted Solingen products. Other Philadelphia sword makers followed the same pattern, most especially Horstmann, whose blades were also German and whose sword room had tools for making hilts and scabbards, but not for etching and grinding blades.[29]

Through the first half of the 19th century, the manufacturer continued to compete with the importer. Large import houses dating from the 1820s continued to flourish, in some cases through the Civil War. Several of these houses merit our attention as examples of the trade and how it worked. Lewis & Tomes was a Birmingham/New York house, with Edward Lewis remaining in England and Francis Tomes arriving in New York in 1819. The firm imported all manner of military goods including guns, swords, accouterments, and buttons. According to the New York city directories the firm broke up in the early 1830s, but at least one pair of post-1850 revolvers has been noted with the Lewis & Tomes mark, indicating that they still served an import trade. The successor firm in the United States was Miller & Tomes, and later Tomes, Melvain & Co., which carried the house through the Civil War. They went bankrupt in 1883.

One importer whom we meet in greater depth is Adam W. Spies. His

[29] "Horstmann Machine Book". Coles Lippincott Horstmann Collection, Historical Society of Pennsylvania. Entry for Horstmann's purchase of Widmann's presses and tools, January, 1849.

relatives donated to the New-York Historical Society a sample book of buttons which Spies had assembled in Birmingham in 1827 (Plate 3). We know that he was born in New York in 1800 and that he joined the firm of J. D. Wolfe which apparently sent him to Europe as its agent. After his return to the United States, he extensively imported arms as well as smaller articles such as buttons.[30] In fact the arms portion of his trade soon overtook any other line of goods. He is thus known to us from having his name on items he imported such as pistols and swords.

Spies was joined in the arms import trade by many other individuals. The great firm of Henry Young & Co. of New York also imported various types of arms and other military trimmings. Dating to 1818, the Young firm was one of the largest houses of its day, becoming Young Smith & Co., then Smith Crane, and ultimately Shannon, Miller & Crane after 1866. Another spin-off was J. Rutsen Schuyler, the elder partner in Schuyler, Hartley & Graham, who got his start with the Young firm.

Another important veteran of the old importing houses was Marcellus Hartley. A biography of Hartley by J. W. H. was issued on Hartley's death in 1903.[31] Hartley began with Tomes, and managed its sporting goods department in St. Louis, Tomes apparently having a connection with another importing firm, Henry Folsom, of that city. Folsom began with H. E. Dimick & Co. in the arms trade. His R. G. Dun and Company rating of Dec 13, 1879, states

> ". . . formerly imported the bulk of his goods, but for some time now has imported little. . . . He came to St Louis as agent for Tomes of N.Y. (entry April 27, 1860) & continued as such until a year past when he commenced on his own account . . . ." [32]

Hartley stayed with Tomes until 1854, when he joined J. Rutsen Schuyler and Malcolm Graham in founding Schuyler, Hartley & Graham.

During the Civil War Hartley used his talents to serve as Stanton's agent in Europe, competing for the purchase of arms with the Confederacy.[33] He seems to have been commissioned Brigadier General to give

[30] *Our Firemen: A History of the New York Fire Departments.* New York: Augustine E. Costello, 1887, pp. 426-433.

[31] J. W. H.: *Marcellus Hartley A Brief Memoir.* (Privately printed, 1903), pp. 51, 52, 56.

[32] Bazelon, Bruce S. & McGuinn, William, *A Directory of American Military Goods Dealers and Makers.* Manassas, VA: REF Publications, 1995. Entry for Folsom.

[33] *Op. cit.* Appendix describing Hartley's work for the Lincoln Administration.

him the necessary authority to negotiate. He also states that he made nothing on these transactions. He does not say what he made when the Europeans needed all their guns back in 1870. Suffice it to mention that Hartley could afford to found the Union Metallic Cartridge Co. and to buy the Remington Arms Co. When he died, Hartley was a very influential figure on Wall Street, meriting an obituary of several pages in the *New York Times.*

All this tells us several things about the merchants of New York in the arms trade. First, it seems that they knew each other and worked together when it was to their benefit. Second, younger generations of talented businessmen served apprenticeships in one of the houses and then perhaps went off on their own. It also helped to have a relative in the business.

Most importantly, the history of these firms shows a shift of emphasis away from import to a system of internal fabrication and distribution as the American economy strengthened. After all, merchants were merchants first and importers second. They would do what was necessary to make a profit. When American manufactured products could compete with imports–and with tariff protection compete favorably–the best of the importers simply switched their trading emphasis to domestic products.

Finally, we now need to turn to the system of internal distribution. Large merchants had to distribute their goods. They could simply sell them as packages upon arrival, or they could develop long-term business relationships with customer firms located in other cities and towns. In some cases they went so far as to set up a branch house in other locations, as did the Folsoms, Benjamin Kittredge, and others by the 1840s and 1850s. This practice became very common later in the century, when most large military goods dealers had branch stores. Stokes Kirk of Philadelphia issued catalogs which mention some 37 branches located all over the United States. They handled most of the government surplus Krag rifles, directly supplying the Alaska trade with them.[34]

Like the merchants, manufacturers also needed to distribute their wares. Some prominent artisans had a national correspondence with their customers. Scraps of this activity have survived. Henry Cressman, a Philadelphia cap maker, wrote to such diverse customers as the State

---

[34] R. Graham Dun papers. Baker Library, Harvard University School of Business Administration. New York vol. 316 p. 166, p. 200 N/275. Entry 12/13/1879 states that Charles Folsom had formerly imported the bulk of his goods but (recently has imported) little or none. "Deals mainly in American Goods, which he buys in the market . . . ."

of Virginia, and to the First Light Infantry of Providence in the 1830s.[35] Cressman was doing a national business in manufacturing the somewhat hard-to-make bellcrown caps of the 1820s and early 1830s. He was displaced when an easier form of cap was adopted in the middle 1830s and makers such as John B. Baker of Boston took over his trade.

Henry Storms and his sons Christian S. and Henry J. Storms represented another aspect of business. His R. G. Dun and Company rating shows him as "always a politician" and therefore not a good business man. However, his politics did get him the post of Commissary General of New York state. His sons continued to run the business, presumably supplying what their father then approved.[36]

Ethics in the last century differed from those of today. It was a normal part of business for military suppliers and makers to be members of the militia companies they accoutered. Often the West Point tailor continued his trade among the cadets he got to know while on post. Regular army officers tended to stick to houses which had a reputation for supplying them, regardless of where they were stationed. A series of letters from John Earle & Co. of Boston dating from the 1850s indicate that officers such as John B. McClellan, G. W. Slocum, Winfield Scott Hancock, and other well-known Civil War personalities were ordering their uniforms from the Earle firm in Boston even though they were stationed all over the United States.[37]

Another firm that traded with the regular Army was Charles St. John of New York, whose 1830s caps enjoyed a wide distribution. John Smith Fraser, whose brother appears to have been a well respected cadet at West Point, certainly started his tailoring business on that post and then moved to New York.[38] He also made the standard samples for the U. S. Army engineer clothing of the 1840s. The Dingees, also of New York, made regular Army accouterments on contract during much of the first

[35] Letters of Henry Cressman. Virginia State Library Reports–Superintendent of the Armory and Commandant of the Public Guard. Box 2 (1836-46). Cressman to Bolling 6 Feb. 1838, 29 Oct. 1836-1846. Letter of Blair Bolling 6 Feb. 1838. Cressman to Bolling 7/12/1828. Anne S. K. Brown Collection. John Hay Library. Brown University. Cressman to William Brown 11/8/1832. Baker to William Brown 11/8/1832. Brown letter 11/8/1832. William Cressman to Brown 7/24/1835, William Cressman 8/18/1836.

[36] *Ibid.* New York vol. 316 a, p. 147, 11/22/33.

[37] Earle, John. Papers in the New-York Historical Society consist of a small number of letters to well-known officers dating from the late 1840s and 1850s.

[38] Samples for Engineer insignia submitted as standard samples by Fraser in the mid-1840s. Other "inclusions" have disappeared. The samples are included in RG-92, National Archives.

half of the 19th century. They also made leather equipment for the militia when not otherwise occupied by Federal contracts.

Anson Mills has a unique story. Born into a poor family in Indiana, he learned the weaving trade as a child. Unlike almost any other artisan, he attended West Point. During his subsequent Army career he worked on his own with the Russell Manufacturing Company of Middletown, Connecticut, to devise a belt with cartridge loops integrally woven into the belt itself. Continuing his Army career and leaving the subsequent business of the Mills Woven Cartridge Belt Company to his brother-in-law, T. C. Orndorff, who received a small percentage for his efforts, Mills proceeded to become what was believed at the time to be the richest officer in the U. S. Army.[39]

The military goods and arms trade managed to make money not only by supplying the troops but also by redistributing used merchandise. As far back as the 1830s, Cressman negotiated to take in trade old hats which he could then resell. Some dealers clearly specialized in buying old muskets and shipping them to other countries, though none as successfully as Marcellus Hartley, who purchased arms for the Union as a public service, but made a fortune selling them back to Europe five years later for the next unpleasantness.

Horstmann's direct competitors in the military trimmings trade in Philadelphia were Henry Korn, Henry Duhring, and Peter and William G. Mintzer. There was also a good deal of competition among the many other textile mills in Kensington and Roxborough, some of which competed with Horstmann in the silk fringe trade. We know the most about Korn. He is probably the individual noted in Filby as having arrived in the United States on August 20, 1813, from Great Britain–in spite of his German name–and is first listed as a lace and fringe weaver in 1814, continuing solely in that trade until 1831.[40] His later listings from 1833 to 1858 are as "military goods". Advertisements from the 1840s show him purveying pompons, tassels, bindings, shoulder straps, sashes, dress aiguillettes, and straps. Like Horstmann, he also sold a general line of swords, buttons, flags, and other military items which he had no capacity to manufacture.

Korn's record in the 1820 Census of Manufactures shows his manufacturing capabilities as utilizing raw materials of wool, cotton, and flax. He consumed about 600 pounds of wool, 1,000 pounds of cotton, and 300

---

[39] Mills, Anson: *My Story* (D.C.: Byron Adams, 1921), pp. 24, 310-327.

[40] Filby, P. William: *Passenger and Immigration Lists Index* (Detroit: Gale Research Co, 1982), p. 196.

pounds of flax, all of which cost him $675. He employed at that time one man, two women, and three boys who were employed in using 10 lace and fringe looms, one warping mill, a gimp mill, a twisting mill, and various other appurtenances of the narrow textile trade. He invested $5,600 in capitalizing the business, and paid out $572 in wages. Because of the difficult times, his product only had a market value of $4,000.

The next available census of manufacturers was taken in 1850. It shows Korn together with Duhring, Mintzer, and another manufacturer, William Grey, making military trimmings of the same type as those produced by Horstmann.

The Military Goods Trade

Peter Mintzer, also in the coach lace, fly net, and military goods trade, is listed from 1818 to 1839 and was followed by William G. Mintzer, who specialized more or less in military goods from 1839 to 1869. Henry Duhring advertised in 1829 as Horstmann's successor at 53 North Third. While we will examine this point in detail later on, there is a check pictured in the Horstmann centennial book dated June 7, 1824, and signed by both Horstmann and Henry Duhring. There is no other indication of any collaboration between the two firms except for this meager evidence.

In any case, the two ran separate courses after 1829. Duhring is believed to have immigrated to the United States in 1818 from Mecklenberg, Germany, and to have begun about 1824 in the lace fringe and military business. His listings in military goods are from 1828 to 1852, although his business continued afterwards through the late 1860.[41]

TABLE 1
Census of 1850

Manufg.	Capital	Male Wkrs.	Female Wkrs.	Materials	Value	Product	Value
Korn	$ 1,500	2	6	worsted,cotton, other	$ 1,000	Military trimmings	$ 3,000
Duhring	$ 10,000	5	20	silk, cotton, other	$ 12,000	Fringe, warps	$ 20,000
Mintzer	$ 27,000	5	35	cotton, linen, thread	$ 11,000	Fringe, cord, tassels	$ 40,000
Wm Grey	$ 100	2	0	silver, brass, other	$ 450	Epaulets	$ 1,500
Horstmann	$200,000	70	130	gold, silver, brass, silk, worsted, cotton, other	$100,000	Epaulets and various military articles	$180,000

[41] *Biographical Encyclopedia of Pennsylvania of the Nineteenth Century*, (Philadelphia: Galaxy Publishing Company, 1874), pp.169-170. Entry for Louis Duhring.

From the above table, we can see that while Horstmann and Korn were probably about equivalent in 1820, by 1850 Horstmann far outdistanced his competitors. His real equivalents were the textile mills of Kensington; Lowell, Massachusetts; Rhode Island; and soon thereafter, Paterson, New Jersey. In the military field, Horstmann's phenomenal growth vis-a-vis his competitors continued. By 1860, only one firm of the above table was still in business along with Horstmann, although Grey would re-emerge during the Civil War as a sword maker. A comparison of Horstmann and William G. Mintzer in 1860 shows that while Mintzer did not even retain equality to his 1850 rating, the value of Horstmann's product had multiplied almost geometrically from $180,000 in 1850 to $600,000 in 1860, a huge capitalization for the times.[42]

TABLE 2
Census of 1860

Manufg.	Capital	Male Wkrs.	Female Wkrs.	Materials	Value	Product	Value
Mintzer	$ 5,500	5	30	cotton yarn, linen, silk, worsted	$ 3,875	Fly nets, cord, tassels, fringes	$ 13,550
Horstmann	$950,000	200	250	silk, cotton, worsted, brass, gold, silver	$300,000	Military goods	$600,000

The changing economic situation in the United States was partially responsible for Horstmann's rise. A brief overview of these changes shows how both colonial and Federal America sorely needed skilled artisans. Colonial theory of the 18th century did not encourage manufactures or trades in the colonies, which were seen as a source of raw materials and a market for finished goods. The close of the American Revolution left the infant United States still very much dependent on European, and especially English, imports. By the turn of the century a patriotic awareness of the importance of fostering domestic industry had resulted in a protective tariff for U. S. manufacturers. While the unstable European political situation also fostered American manufactures, the United States in 1790 continued to be largely dependent upon foreign products.

[42] *Manufacturers Census of the United States*. Schedule E. Philadelphia County. 1850 and 1860. Entries for Horstmann.

Through the first half of the 19th century, this situation gradually but nonetheless dramatically changed. By 1850 the United States was to a great extent producing what it consumed. European imports were valued to an increasing extent because of their snob appeal. The catalysts of this change were artisans like Horstmann who brought badly needed skills into a growing market. This market in turn produced surplus capital in which industrialization was rooted. Unlike many immigrants, Horstmann seems to have had both the capital and the industrial expertise to strike out on his own in a major way. However, as we have seen, he was by no means the only person to arrive in America and set himself up in the lace trade or in the military manufacturing business.

5th and Cherry Street factory, early 20th century view

Fifth and Cherry Street factory showing original building c. 1859 and new sales rooms added during the Civil War. Horstmann scrapbook.

Chapter 3. Horstmann in Philadelphia

Early 19th century Philadelphia, with its German population and its newly important textile industry, was a likely destination for a man like Horstmann, and was well suited to his talents. No longer young and with a varied background which included both a mastery of a skilled trade and experience in a textile factory, Horstmann's experience was well suited to his adopted country.

The exact date of Horstmann's arrival is not known. The immigration records of the time include only his declaration of intent to become a U. S. Citizen, dated October 13, 1817, and, later, his oath of allegiance on June 10, 1823.[43] Horstmann arrived in Philadelphia probably in late 1815 or early 1816. The article, "Lace Making in the United States," which appears among the Horstmann papers, gives the date of May, 1815, the earliest citation found. This should be contrasted with the R. G. Dun and Company report that Horstmann was at the Battle of Waterloo in June of 1815.[44] The confusion extends back to the middle years of the 19th century, when the company's founding date was given variously as 1815 or 1816 during the lifetimes of Horstmann's immediate family. The firm's official biography (1916) states that Horstmann witnessed the downfall of Napoleon and sailed for Philadelphia early in 1816, and this

[43] Filby, P. William: *Passenger and Immigration Lists Index*. (Detroit: Gale Research Co, 1982), p. 293.

[44] R. Graham Dun Papers, Baker Library, Pennsylvania vol. 132, p. 511, entry Oct. 6, 1850.

is probably as nearly accurate as can be discerned in the absence of other records.[45]

William H. Horstmann apparently found a congenial atmosphere in Philadelphia's Germantown area quickly after his arrival. Germantown was then a separate community located just north of Philadelphia. Although there was a German community in the area by the 1680s, that community was quickly losing any exclusive German character it might have had, while still retaining a strong German presence.

Horstmann was apparently first employed by Frederick Hoeckley, a well-known lace and fringe weaver in Germantown.[46] Given that Horstmann appears on his own in the 1817 Philadelphia directory, his employment with Hoeckley must have been of a temporary nature. As a more permanent relationship to the Hoeckleys, he married Sarah Hoeckley, the daughter of his former employer, in 1816 or early 1817. Marriage records in Philadelphia's Lutheran Archives fail to record Horstmann's marriage either in Germantown or in either of the two Lutheran congregations in Philadelphia itself. In the absence of these records, the birth of a first child in November of 1817 sets a reasonable date for the marriage earlier that year. The connection with the Hoeckley family is probably not a coincidence. Frederick Hoeckley was born in Elsingen, Germany, in 1758.[47] While he may be the same as a Jacob Frederick Hoeckele known to have immigrated in 1771,[48] Hoeckley cannot be positively identified in Philadelphia before the early 1790s. Hoeckley was in Philadelphia in October, 1792, when his daughter Sarah, the future Mrs. Horstmann, was born in Germantown. Hoeckley married Anna Gottlieben, whose family was in Philadelphia from at least 1727.[49]

Hoeckley signed a number of wills as witness, and was evidently quite well known in the German community which centered around St. Michael's Lutheran Church in Germantown.[50] By marrying Sarah

---

[45] *One Hundred Years*, p. 13.

[46] *One Hundred Years*, p. 14.

[47] Frederick's life dates and the birth date of Horstmann's daughter, Henrietta, are taken from their tombstones at St. Michael's Lutheran Church in Germantown, as is the place of Hoeckley's birth.

[48] Filby, P. William: *Passenger and Immigration Lists Index.* 1988 Supplement. (Detroit: Gale Research Co., 1988), p. 18.

[49] *Collections of the Genealogical Society of PA*, vol. 8, Phila., 1896.

[50] *Ibid.*, Phila. County Will Book 7, p. 91, 8, p. 123.

Hoeckley, Horstmann aligned himself with an established German trades family which was quite well known in the Philadelphia area. However, as Sarah is noted in Frederick Hoeckley's will as only one of his twelve children, it is unlikely that Hoeckley could contribute much of a financial start to Horstmann's incipient business.[51] In spite of their prominence within the German community, neither Hoeckley nor Horstmann can be documented as having been overtly religious. Not only is Horstmann's marriage unrecorded in the surviving Philadelphia Lutheran church records, Hoeckley and Horstmann fail to appear in the lists of communicants carefully prepared each year by St. Michael's in Germantown.[52] In fact, Horstmann's first known connection with the church is the burial of his daughter there in 1838.

Hoeckley is credited by the lace trade as the original lace weaver in the Philadelphia area. An article entitled, "Lace Making in the United States", among the Horstmann papers at the Historical Society of Pennsylvania, cites Hoeckley's advertisement of May 25, 1793, as a fringe and lace maker.[53] The advertisement states in both English and German that Hoeckley has removed his business from Germantown to 322 North Second Street in the Northern Liberties section of Philadelphia, where he "makes all sorts of fringed and plain laces, livery laces . . . and all sorts of fringes . . . ." Hoeckley appears in the Philadelphia City Directories after this move to 322 north Second Street in 1793. By 1800 Hoeckley was at 304 north Second where he remained until after 1819.

Horstmann, whether with his own funds or borrowed money, was soon in a position to set up his own business. The 1817 Philadelphia City Directory mis-spells his original entry as "William Hortman" at 12th and Arch Streets. The next year, Horstmann had moved to North 3rd Street, and was listed as a fringe weaver. His listings continue in that trade through 1828, when the business was first listed as "Military Store". This is the first absolute indication of Horstmann's dealing in military equipments and ornament. As such he joined competitors Henry Korn and Henry Duhring in making woven military fringes as well as civilian ornament and such well established businesses as Henry Cressman's

[51] *Collections of the Genealogical Society of PA*, vol. 80, 1907, will dated July 21, 1823.

[52] Now at the Lutheran Archives Center, Krauth Memorial Library, 7301 Germantown Avenue, Philadelphia.

[53] The advertisement is printed in an article on Lace Making in the United States from an unknown journal but among the Horstmann scrap book papers at the Historical Society of Pennsylvania.

military cap manufacturing establishment, which had developed a national clientele.

Henry Duhring appears to have been in partnership with Horstmann during the middle 1820s. Unfortunately no city directories were published for several years during this time and we have to turn to other sources for confirmation in the absence of any statement at all on the subject in the 1916 history of the firm. However, a check dated June 7, 1824, which is signed by both Horstmann and Duhring is pictured in the 1916 history. Further corroboration appears to be supplied by the R. G. Dun and Company Collection Philadelphia ledger 1 page 1 which gives Duhring's June 4, 1856, listing as "Was once a partner of W W Horsten (sic) & made money in that connection. Has been in bus 20 yrs. Imports most of the goods he sells, and has been regarded as rich for many yrs . . . ." R. G. Dun and Company's actual estimate of Duhring's wealth in its April 12, 1864, entry is $750,000, so indeed he had done well.

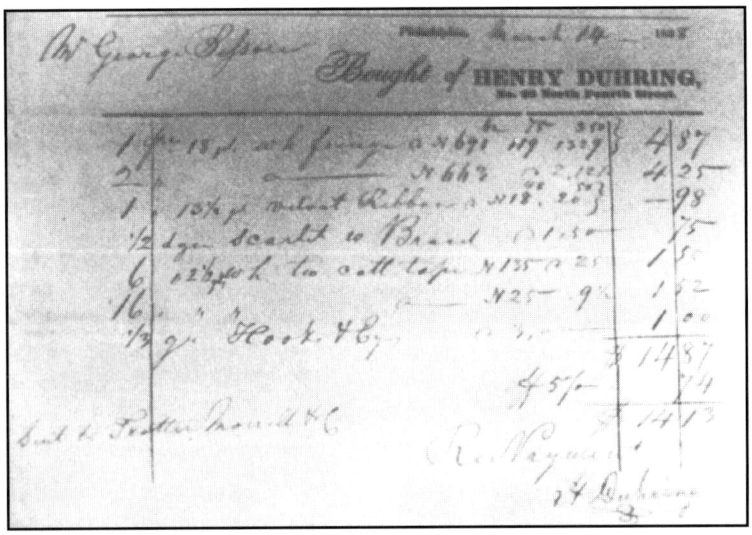
Bought of HENRY DUHRING,

In 1829 Duhring advertised as Horstmann's successor at 53 N Third. It is significant that the years of Horstmann's probable association with Duhring coincide with the times he was journeying to Europe in search of machinery and hoping to set up his factory. In this context, an association with Duhring explains who took care of Horstmann's business interests while he was abroad and reinforces the inception of the import and manufacturing duality which characterized the Horstmann business through the remainder of the century. Duhring was an importer of military goods from 1828 to at least 1852 and his association with Horst-mann may also have prompted the latter to become heavily involved in the military goods trade

On the other hand, there is evidence against there ever having been a firm entitled Horstmann and Duhring, as the custom of the times would have found usual. Not only has no artifact yet been seen with such a label but a small strap work box dating from this era is labeled "Manufactured by George Horstmann at Essen Cassel for W. H. Horstmann & Co. 53 North 3d at Philadelphia". Inside was once a No 10 gold curl ("1 Schfl Oval Kraus N 10 7 6"), probably a rolled length of gold trim. It is likely Horstmann and Duhring had some sort of business arrangement which did not compromise their independence as separate entities. In the process, Horstmann may well have learned some aspects of the military goods import trade from Duhring who specialized in this business whereas there is no provable military manufacture or transaction dating from Horstmann's first years.

The Horstmann business clearly grew rapidly through the 1820s and 1830s. Horstmann's city directory listings show him renting first at 51 North Third, and then as having purchased an entire store at 55 North Third, which in later years he completely rebuilt. His listings show him as a fringe maker through the middle 1820s, but when the listings resume in 1828 after a two year hiatus, Horstmann's business appears as "military store" at 48 North Third. This listing represents Horstmann's first certain connection with the military trade. These listings are amplified by a statement made in *One Hundred Years* to the effect that "When William H. Horstmann commenced business in Philadelphia the articles he made were chiefly those described as proper to the passementier's art. He gradually took up the manufacture of military equipments, such as pompons, epaulettes, hatbands and similar goods. Military wares and manufactures begin to assume importance from about the year 1828 . . . ."[54]

The Horstmann firm was apparently soon thereafter dealing in all forms of military goods and weapons, excepting guns. This bias reflected the fact that muskets were issued to, rather than bought by, the militia. While officers purchased their own weapons, there was plenty of competition in the gun trade, and Horstmann never entered that side of the business. Horstmann's firm, in fact, retained a close connection to its founder's origins in the lace trade. The civilian portion of the business, making coach lace and later thread, would continue to be a large and viable part of the firm well into the 20th century. In the 30 years during which Horstmann actively managed his firm, its manufacturing capacities seemed to have been limited to forms of lace and woven ornament,

---

[54] *One Hundred Years*, p. 25. (*The Philadelphia Gazette and Daily Advertiser*, Nov. 6, 1827, p. 2, col. 4, notes the brig *Veteran* from Hamburg carrying "sabres, six cases, pictures, one case . . ." to W. H. Hortsman [sic]. This is the first known instance of Horstmann in the military goods trade.

even though the business bought and sold all sorts of other military items.[55]

Horstmann Senior is believed to have made at least two trips back to Europe during his active tenure with his firm. These are cited in various accounts of which the earliest is Freedley, writing in 1859.[56] These trips apparently resulted in Horstmann obtaining plaiting and braiding machines for his Philadelphia factories. In 1838, Horstmann took his eldest son, William J. Horstmann, with him leaving the management of the business to his teenage second son, Sigmund. The firm's scrapbook contains several newspaper articles which mention Horstmann's first return visit to Europe as being in 1823 or 1824 followed by the importation of machinery the next year. The clearest of these is in a scrapbook article dated 10/17/1874.

Jacquard machines were apparently imported by the Horstmann firm in 1825 from Continental Europe. Where the imported machinery can be identified, these tools came primarily from Germany, France, and Switzerland. This is supported by a notation in *One Hundred Years* that Horstmann was shipwrecked off Le Havre, France, in 1824, and performed bravely in helping to save others less athletic than he was. The machinery that is noted in the bill of sale for the auction of Horstmann's Columbia Manufactory in 1853 also cites a number of the machines as French or Swiss. One source, Freedley, states that at least some of the machines, singling out the plaiting and braiding machines, were imported from Germany.[57] Significantly, no source gives any hint of a connection between Horstmann and the large British textile industry in any business enterprise. For whatever reason, Horstmann's focus and sources of trade remained exclusively Continental, in all probability reflecting the preferences and prejudices of Horstmann's youth and training on the Continent.

France at this time was the great center of the continental textile trade and as such did a thriving business with such countries as the

---

[55] The Horstmann Machinist Book in the HSP collections covers all machinery purchased by the firm from 1842 to 1870. There is no indication before 1845 of any capabilities in manufacturing of metal ornament, swords, drums, or flags. This is also reinforced by the auction sale handbill from 1853, when the contents of Horstmann's factory on Columbia Avenue were sold. This handbill lists only the textile equipment in which the Columbia Manufactory specialized.

[56] Freedley, Edwin T: *Philadelphia and its Manufactures: a Hand-Book* (Philadelphia: Edward Young, 1858 p. 246), *One Hundred Years* p. 15. The former source gives the date of the first trip as 1824, the latter, 1823.

[57] Freedley, 1858 edition, p. 249.

United States. In the mid years of the 19th century the Horstmann firm maintained a branch office in Paris, apparently for buying silk and other textile items for resale. As noted above, much of the woven military ornament imported into the United States which can be ascribed to a country of origin seems to have originated through Le Havre from Paris and Lyons.[58] Horstmann's own sample book of labels features examples intended for French goods in the 1850s.

Horstmann in Philadelphia

Although some of Horstmann's training may have been French, his trade loyalties apparently continued to lie with Germany, at least in the firm's first decades. Horstmann's first known import was in February, 1819, when the *Milo* arrived from Hamburg. Aboard was a box of merchandise consigned to William H. Horstmann & Co. by Menge & Nieman and another arrival dated on September 14, 1819.[59] The following table gives known imports by Horstmann. With one exception they all originated from the German port of Hamburg.

TABLE 3
Horstmann Imports 1821-1826

Date Arrived	Ship Name	Imports	Consigned to	Departed from
2/19/1821[60]	*Aeolus*	1 case merchandise, 1 parcel samples	W. H. Horstmann	Hamburg
2/9/1823	*Hunter*	5 cases	W. H. Horstmann	Hamburg
3/23/1824	*Daphne*	1 box Mdze	W. H. Horstmaner	Hamburg
7/2/1824	*Maria Elizabeth*	1 package mdze.	W. H. Horstman	Hamburg
7/16/1824	*Polydore*	11 cases & 1 bbl mdze	W. H. Horstman	Hamburg
9/1/1824	*Hibernian*	18 cases, 1 bale	W. H. Horstmann	Hamburg
9/13/1824	*Adams*	1 chest	W. H. Horstman	Amsterdam
10/23/1824	*Delight*	1 bale, 3 cases	W. H. Hortsman	Hamburg
12/6/1824	*Hermima*	7 cases, 1 parcel	W Horstman & Co	Hamburg

[58] *Columbian Centenel*, Boston May 4, 1826, p. 4 col. 3, gives a clear indication of John J. Low importing metal goods from England and woven items from Paris.

[59] U. S. Bureau of Customs, National Archives. Records for the *Milo*, arrived in Philadelphia February 22, 1819, and the *Hibernia* which arrived September 14, 1819, each with a case consigned to Horstmann.

2/21/1825	*Vesta*	Mdze 2 boxes	W. H. Horstman	Hamburg
6/12/1825	*Hibernia*	Mdze 10 cases	W. H. Hasturan	Hamburg
9/30/1825	*Plato*	Mdze 10 cases, 2 trunks, 1 box, 18 chests	W H Horstman	Hamburg
6/7/1826	*Neptune*	13 packages, 30 bbls, 2 cases, 1 parcel	W Hartman & Co	Hamburg
9/18/1826	*Sarah Ralston*	10 boxes	Wm Hortsman	Hamburg

Aside from the fact that the persons transcribing the ships' manifests into newspaper information could not spell particularly well, the above table does show that Horstmann shipped the overwhelming majority of his imported merchandise aboard ships leaving Hamburg for Philadelphia. Indeed, only three vessels were noted making this voyage during this time period which did not carry a consignment for W. H. Horstmann. Only the single vessel from Amsterdam carried a shipment for Horstmann which did not originate in Hamburg. Horstmann was not importing on the ships noted from Liverpool, which were frequent, or on the comparatively rare vessels from Bordeaux or LeHavre. Excepting only the first entry, the information is taken from the "Imports" column of the "United States Gazette," a new feature in the summer of 1823. The first citation, from the U. S. Bureau of Customs manifests, shows the consigner as a Berudis Diestraum (sic) with case marks of HWH/ GO/ WH on the packages. Unfortunately, the newspaper seems to list fewer vessels than are known to have made port and cannot be considered inclusive. Even with the record as presently exists, the overwhelming German character of Horstmann's overseas business connections during the 1820s can be ascertained.

The Census of the Horstmann machinery taken in February of 1845, when Horstmann left the firm, notes many machines as having been imported from France, some shortly before that date. In one case, an entry in the machine book just subsequent to the inventory in August of 1845 shows six Jacquard machines, 100 hooks imported through J. Bouge, Paris, for what appears to be $1,109.[61] One other indication of Horstmann's early importing interests can be ascertained from a small wooden box with a label reading "manufactured by George Horstmann

[60] The entries for the *Aeolus* and *Dido* are drawn from the records of the U. S. Bureau of Customs, National Archives.

[61] Coles Lippincott Horstmann Papers, Historical Society of Pennsylvania. Machine Book entry, August 9, 1845.

at Hessen Cassel for Wm. H. Horstmann & Company, 53 North 3rd at Philadelphia". On the side of the box in pen is a note of its former contents, "a golden curl" (translated from the German). It would appear from this evidence that Horstmann was also importing finished products from his family in Cassel.

The Jacquard loom, of which Horstmann is claimed to have been the original importer to the United States, functioned as a mechanical computer, complete with sometimes thousands of punch cards which controlled the patterns woven by the loom and enabled it to weave the same complicated designs time after time. Although prototypes seem to have antedated Jacquard's original design, the new system is named after Joseph Marie Jacquard, a native of Lyon, France, who employed it about the year 1801. The Jacquard loom revolutionized the textile trade much the way the cotton gin facilitated the production of cotton.[62]

It was these Jacquard machines which Horstmann introduced into the American lace trade. He is credited in an article in the Horstmann scrapbook produced by the lace trade about 1900 as being the first to introduce the Jacquard loom itself. This is a difficult matter to prove and in fact Horstmann was simply one of the early manufacturers who employed the Jacquard loom rather than its progenitor in the United States. However, he may indeed have been the first to have employed it in the manufacture of narrow lace. Be this as it may, what is important is that Horstmann was quite familiar with the latest technological advances in Europe. He considered modern technology an important ingredient in the setup of his Philadelphia manufacturing capability and by the mid 1820s had the means to purchase and utilize this advanced machinery.

By 1831 Horstmann had outgrown his manufacturing quarters, which were apparently located in his own shop at 55 North Third, and he set up a factory at Germantown Road and Columbia Avenue in Kensington, a suburb of Philadelphia. This factory was a four story brick building with 20 windows on each floor of the Columbia Avenue side. Its appearance is known to us from the bill of sale cited earlier, when the contents of its machinery were sold at auction upon its closing, and from various bill heads and trade cards issued by the Horstmann company before 1853. The auction closing the Columbia factory was held on September 21, 1853, after which date the Horstmann firm centralized its manufacture and sales in the center city area.[63]

---

[62] *One Hundred Years*, p.15.

[63] *One Hundred Years*, p. 15 and *Op. cit.*, Horstmann scrapbook.

This Columbia Manufactory, as Horstmann's textile weaving facility was known at the time, sold 167 lots of machinery for a little over $250. Some items apparently did not sell. These give us a unique glance into the trade between 1830 and 1850. Included in the sale were 12 fringe looms, a fringe mill, 13 power looms or engines, 22 Jacquard machines of 104, 416, or 800 hooks, alternatively 60 to 80 needles and cards, and a new French quill machine. There were also 17 braiding machines, two plaiting frames, 18 French card machines, 6 lathes with shuttles, 4 fringe loom lathes, two French gimp mills, one machine for drawing gold and silver wire, three tables for making fly nets, one machine for putting up ribbons, a machine for twisting silk, another for sizing silks and velvets, and yet another for putting up braids and cords, an embossing machine with roller for velvet ribbons, and two for daguerreotype cases. The latter brought $28, the high for the sale. While the initial Columbia factory was gone, its name lingered in Horstmann's product line till the last days of the firm in its "Columbia yarns" and trademark of a "shepardess", who is in fact the personification "Columbia" wearing the traditional Phrygian cap of this American icon of liberty.

An inventory taken of Horstmann's machinery on February 1, 1845, at his retirement, shows again that his manufacturing interests were clearly in the fringe and lace industry. This inventory lists such items as 55 looms, including engine, coach lace, seaming fringe, and lace looms; 18 Jacquard looms, of which 10 were from Paris; 64 braid machines, of which 18 are listed as being from Paris, two from Berlin, and several others from vendors with German names. There are also 18 twisting wheels, 19 spooling wheels, 46 swifts, four gimp mills, one from Lyon, ten sash frames, and many other shuttles, bags, peels, quills, etc. valued at $2,971.54.

Most of this machinery relates to Horstmann's manufacture of coach lace, fringe, and tassels. These items show the civilian side of Horstmann's business, with probably only a limited application for producing the epaulets and gold/silver wire embroidery which would have been the mainstay of his military manufacture in his early years. In fact, before all of the Horstmann operations were consolidated in the mid-1850s, this military embroidery operation was probably undertaken at Horstmann's downtown Philadelphia location. The dichotomy of military and civilian production would go on throughout the entire life of the firm, to be reflected in later years by two parallel firms within the Horstmann Company, and later as separate departments within a consolidated firm, and ultimately as two separate firms operating within the same factory location.

The Horstmann family was also growing, in much the same fashion as his business. Horstmann married Sarah Hoeckley, the daughter of his apparent former employer, in 1816 or early 1817. His first child was Henrietta G. Horstmann, born November 17, 1817.[64] His next children assured the survival and continuation of his business. William J. Horstmann was born October 7, 1818, and Sigmund H. Horstmann July 23, 1821. In later years there was a daughter, Josephine, described in the records of St Michael's Lutheran Church in Germantown as having died in 1838 "about age 12," and possibly another son, Henry, who, if he lived to maturity, did not join the firm and is not mentioned elsewhere in the family annals.

Much of the remainder of this work will deal with Horstmann's two sons and their children. The younger Horstmanns are cited in the Centennial History as having been educated "at a Moravian school near Lancaster, Pennsylvania", an outgrowth of the Moravian establishment at Lititz. The published catalog of students who attended John Beck's Academy for Boys in Lititz includes an entry for William H. Horstmann, which must refer to William J. Horstmann, as having entered the school in 1826, and then lists Sigmund Horstmann as a student beginning in 1828. A "Harry Horstmann" from Philadelphia is listed as a student beginning in 1842. They would have stayed in Lititz for about seven years.

Lititz is a small community located approximately 75 miles west of Philadelphia, and existed as a Moravian settlement from the mid-18th century. John Beck describes his own career in the same pamphlet which lists his catalog of students. Beck began as a shoemaker's apprentice, but by 1815 he was accepting students from the local community, and a few years later had enlarged his school to accept boarding students, largely from Maryland and Pennsylvania. Many students in Beck's school came from the Philadelphia area. Many but not all of these students have German surnames. Regardless of the students' parentage, German as well as English was taught in the school as part of the standard curriculum. Beck, however, was very innovative for his time and introduced such novelties as separating the older students from the younger, creating a playground, and introducing evening lectures on scientific subjects. Apparently the out-of-town students at Beck's school boarded with local families and were taught a trade such as weaving, barrel-making, carpentry, etc. through an apprenticeship. That Beck was well thought of can be surmised from the number of offers which he had to teach at other institutions or to head up the Moravian education program in

[64] Inscription on her tombstone at Saint Michael's Cemetery, Germantown.

Bethlehem. Beck continued his efforts until his retirement in 1865, after which his school continued for a time, run by others.[65]

In all probability, Horstmann chose to send his children to Beck's school because their education would preserve their German heritage while at the same time giving them a practical and forward thinking course of study. Horstmann was also careful to include William J. on his 1838 business trip to Europe, and to give him a sense of his German heritage. The fruits of this training appeared later on in the continued affiliation of Horstmann's sons with German culture and workmen.[66]

By 1831 the main tenets of the Horstmann business had been set up. The firm maintained an extensive manufacturing capability along with a direct sales room to the public. This sales room handled merchandise that the company made as well as other items which they either purchased or imported. Horstmann was also engaged in the military goods trade on an extensive basis after the middle 1820s. The firm was apparently consulted about the new Army regulations of 1832, a practice which was common at the time. Symbolizing this close relationship, the firm printed copies of those regulations in 1837. A copy of this work is now in the collections of the U. S. Army Quartermaster Museum. A bill to the Washington Grays of Philadelphia dated April 16, 1832, shows a drawing of an epaulette and describes the Horstmann business as "Manufacturers of Coach Lace, Fringe, Cord, Braid, Watch-Guards, Hatbands, Tassels &c &c". Above this is the salutation "MILITARY ARTICLES OF EVERY DESCRIPTION". The bill is for white cotton web and buckles.

Horstmann advertised in Huddy and Duval's *United States Military Magazine* about 1840 as, "William H. Horstman, no. 51 North Third St., Philadelphia, and William H. Horstman and Company, no. 8 Maiden Lane, N. Y., Manufacturers and Importers of Military Trimmings. Officers of the Army and Navy Equipped according to Regulation. Volunteer companies supplied with every article." In addition, Horstmann is known to have dealt in epaulets, swords, caps, and other military equipments and uniform items.

Another statement of Horstmann's business capabilities in the 1830s appears in *The Philadelphia Circulating Business Directory for 1838.*[67]

---

[65] Beck, John: *Complete Catalog of the Names of All the Students who have Attended the Litiz Academy for Boys from A.D. 1815 to A.D. 1865*, pp. 39, 42.

[66] *One Hundred Years*, p. 16.

[67] Savale, J. R. *The Philadelphia Circulating Business Directory for 1838* (Philadelphia: Morris's Xylographic Press, 1838), p. 159.

This source shows Horstmann as "Manufacturer and Importer of Epaulettes, Wings, Contre Straps, Pompons, Sashes, Hatbands, Swords, Sword Knots, Military Buttons, Plumes and Caps. Also, Watch Guards, Braids, Coach Lace, Cotton and Worsted Fringes &c. 51 N 3." This statement is important as contrasted with the 1832 billhead. The Horstmanns' dealing in swords and buttons is now important enough to receive separate emphasis.

Philadelphia is important in the history of U. S. military contracting because it was the headquarters of the U. S. Army Commissary General of Purchases who, in the years of Horstmann's career in America, was Callender Irvine. While the Army attempted to diversify its contracting business among various producers nationwide, the firms in Philadelphia were still in a privileged position because of their proximity to influence decisions and to know the latest thinking about military fashions. The regulations of 1832 specified new white cotton web epaulets and other military ornament which were well within Horstmann's manufacturing capabilities, and with this, in 1833, came the first of a great many United States Army contracts.[68]

The published contracts for 1833 cite three contracts, on March 23, July 20, and August 7, requiring the delivery of approximately 700 epaulets for noncommissioned officers (which would have used nonmetallic thread), 2400 shoulder straps, 75 aiguillettes, 165 yards of half-inch gilt Prussia lace, 1,400 yards of white lace one and a half inches, 1000 yards of scarlet lace one and a half inches, 7500 yards of yellow half-inch binding, and 8000 yards of white half-inch binding, as well as other similar items. Other orders were contracted in 1834.[69] These initial orders would be followed by other contracts in 1836 through '38 and 1840-43 for similar types of items. However, Horstmann was by no means the only supplier of these items, even in Philadelphia. Robert Gethen had a contract for bindings during the early and mid 1840s, and Henry Korn also produced tassels and pompons during the 1840s. Korn's business is also described in the 1838 *Philadelphia Circulating Business Directory.*[70] Korn is described there as "Military Trimmings, Caps, Swords, Belts, Epaulets, and all Other Kinds of Military Equipments Suitable for Militia Officers, Infantry, Artillery, Cavalry Troop and Rifle Volunteer Com-

[68] *House Document 99, serial 256, 23rd Congress, First Session.*

[69] *House Document 142, serial 274, 23rd Congress, Second Session.*

[70] Savale, J. R. *The Philadelphia Circulating Business Directory for 1838* (Philadelphia: Morris's Xylographic Press, 1838), p. 159.

panies, Manufactured and Sold Wholesale and Retail, by Henry Korn, No 22 North Third Street." However, even though Korn made and dealt in virtually the same merchandise as Horstmann, by the 1840s, neither Korn nor any of Horstmann's competitors were as large or diversified as the Horstmann firm had now become.

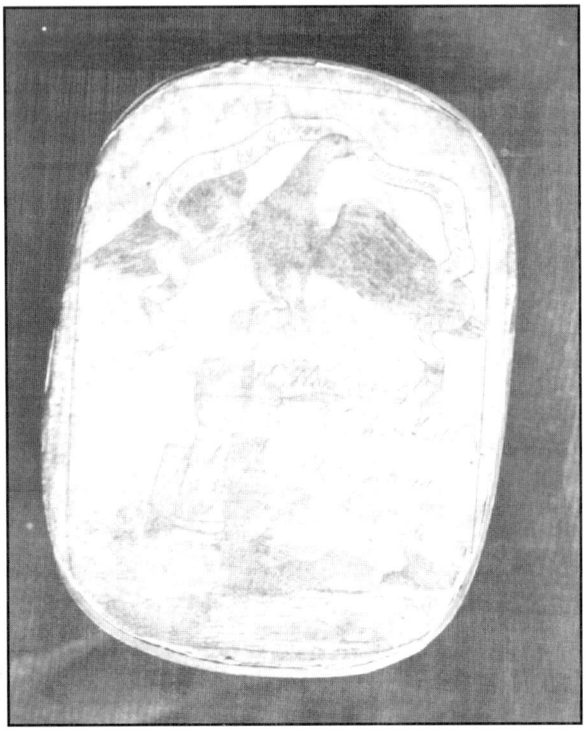

Strap work box with label showing George Horstmann in Hessen-Cassel manufacturing and shipping "a gold curl" to William H. Horstmann in Philadelphia c. 1825. (Author's collection)

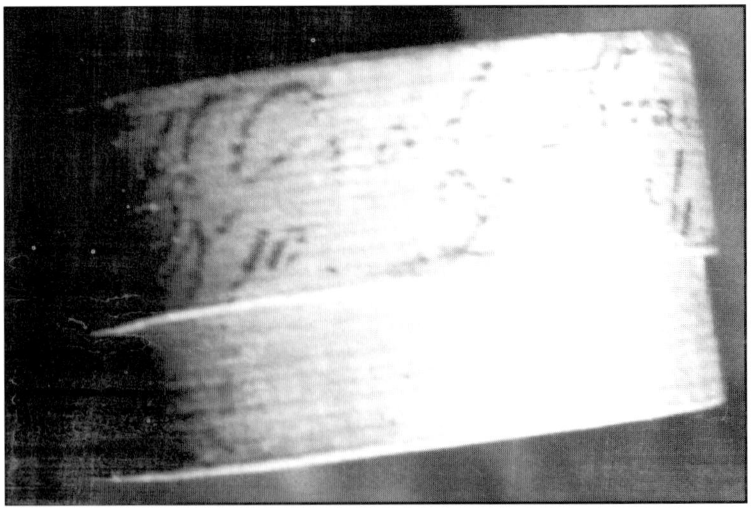

Chapter 4. The Horstmann Brothers

In 1845 William H. Horstmann had been in business in Philadelphia for 30 years. He had trained his sons in his business to the extent that in 1838 when he and his eldest son, William J., had gone to Europe, Horstmann had left the management of his entire concern to Sigmund H. Horstmann, then just 18. While William J. had shown a real understanding of the manufacturing side of the business and had been running the factory in Kensington virtually on his own, Sigmund was handling the mercantile decisions of the firm. William J. also was managing his own outlet for ladies dress trimmings at 204 Chestnut Street.

The Horstmann centennial history states that neither son was taking a salary at this time, but simply drew what he needed from the business. By all accounts, the brothers were mutually complimentary in their skills, highly reliable, and close working partners.[71] This close working relationship would characterize the brothers' method of operation through the next thirty years.

With this pool of talent and trust, William H. Horstmann, now 60, could at least nominally retire from his business. The retirement date was reported in the Philadelphia *Public Ledger* and the *Inquirer*, as February 1, 1845. However, its true nature is shown by the fact that these two clippings from the Horstmann scrapbook are dated June 28, 1848, meaning that it took over three years to get around to announcing the

[71] *One Hundred Years*, p. 17.

elder Horstmann's retirement publicly. In addition, Horstmann senior was listed as a partner in the New York branch of Horstmann Sons & Drucker from its inception in 1845 through its dissolution March 25, 1850. The best indication that the actual retirement was not a fiction is the first page of the company machine book which is an inventory of "machinery as taken from W. H. H. Feby 1st 1845", the date given in the newspapers.

The "boys" were first taken into partnership as William H. Horstmann & Sons in January, 1843 and the new firm is first listed in the 1844 *Philadelphia City Directory*. *One Hundred Years* describes Horstmann's retirement thus: "He could not give up work entirely, however, and accordingly came to work daily and carried on . . . active superintendence of the interests he had founded."[72]

More space was needed for the firm's manufacturing and retail interests. In 1842 no. 55 North Third Street was torn down and a new building was erected on the site, having a sale store on the ground floor, with the upper stories used for manufacturing. An additional building of three stories was put up, which again added to the capacity of both the sales rooms and the factory.[73] Even this growing complex of buildings was insufficient and a new branch was opened up at 723 Chestnut Street "for the sale and distribution of all merchandise".[74] *Rae's Philadelphia Panorama* shows Chestnut Street as it existed in 1851, including W. J. Horstmann's Ladies Dress Trimming store, now at 204 Chestnut [75]

The Mexican War gave the Horstmanns an opportunity to be of use to their adopted country. During Horstmann Senior's direct management of the firm, his adopted country had been at peace and the military mainstays of the Horstmann business were divided between the manufacture of trimmings and the sale of military equipment, basically for the volunteer militia and a small regular Army. Beginning with the Mexican War in 1846-7, the military aspect of the Horstmann firm would become more and more lucrative as the heightened patriotism occasioned by the war translated into the creation of larger and more numerous military organizations during the 1850s.

The Horstmann centennial history, *One Hundred Years,* states that

---

[72] *One Hundred Years*, p. 21.

[73] *One Hundred Years*, p. 19.

[74] *One Hundred Years*, p. 22.

[75] *Rae's Philadelphia Pictorial Panorama Directory and Advertiser* (Phila.: Julio H. Ray, 1851).

Horstmann's patriotism was manifest in equipping a company which it calls the Steuben Rifles, and which may be the Stuben Fusileers cited in Clarke's *History of the Pennsylvania Militia and National Guard* as one of the companies which offered its services for the Mexican War.[76] Neither this company nor the Steuben Rifles appears in the Commonwealth of Pennsylvania's Officer Commission Records which is arranged as a militia "order of battle" for Pennsylvania. They similarly do not appear in the published lists of militia companies whose service was accepted by Pennsylvania during the Mexican War.[77] The officers of the Stuben Fusileers, as given in Clarke, have German names such as Arnold Seyberg, Charles Angroth, and Jacob Schoenlabor. Clarke adds, however, that when the services of this company were not accepted by Pennsylvania, that they joined one of the ten new regiments being formed for the war by the U.S. Army.[78] This assertion is partially confirmed by the record in Heitman that Arnold Syberg served as captain of a company of the 16th U.S. Infantry from February 1847 thru August 1848. He is described as having been born in Germany and serving from Pennsylvania.[79]

Angroth and Schoenlabor have single year entries in the Philadelphia City Directories in the mid 1840s. Charles Angroth actually served in Mexico as a private with the Washington Light Infantry which was accepted as Company E 1st Regiment Pennsylvania Volunteers. This mixed record leaves open how many, if any, of the 78 man Stuben Fusileers accompanied Syberg to Mexico.[80]

The Steuben Rifles, whoever they were, apparently sent a trophy to Horstmann consisting of Mexican saddles mounted on silver, a general's coat richly embroidered in gold, the spare cork leg of Santa Anna, complete with boot and spur, as well as other souvenirs of the war, which at one time were proudly displayed within the Horstmann Company.[81]

---

[76] *One Hundred Years,* p. 26. Clarke, William P.: *History of the Militia and the National Guard of the State of Pennsylvania* (Charles J Hendler, 1912), Vol. II, pp. 46-47.

[77] Records of the Bureau of Accounts, Auditor General. Military Accounts, Mexican War Box 1. RG-2 Pennsylvania State Archives.

[78] *Ibid.* Clarke, Volume II, p. 47.

[79] Heitman, Francis B: *Historical Register and Dictionary of the United States Army* (Washington: Government Printing Office, 1903), Vol. 1. Entry for Arnold Syberg.

[80] Records of the Bureau of Accounts, Auditor General. Military Accounts. Mexican War. Box 1 RG-2, Pennsylvania State Archives.

[81] *One Hundred Years,* p. 26.

Horstmann continued his semi-retirement through the late 1840s. He is cited as planning another trip abroad, but this was not to happen. While he finally began to take some leisure time, his health was failing. He and Sarah went to Bethlehem, Pennsylvania, to recover his health in the sympathetic German surrounding of the Moravian community. The Moravians were justly renowned for their music.[82] The Horstmann firm history mentions that Horstmann loved music and favored the artists of the day when they came to Philadelphia. These included Ole Bull, the Norwegian-American musician who was greatly celebrated in the mid-19th century, and a daughter of the composer Spohr. However, the music and the surroundings of Bethlehem did not cure William H. Horstmann. He died there, August 5, 1850.

The personality of the senior Horstmann, so far as it can be known today, is best summed up in *One Hundred Years.* He is pictured there in his retirement coming regularly to work and studying all aspects of the business he had founded, from the setting up of the looms to the packaging of the finished goods. While there is no way to prove the accuracy of this description, it is consistent with the personality of the craftsman-technician which Horstmann seems to have been. He spent his lifetime carefully biding and tending his business. While capable of bold business decisions involving some risk, in fact he never strayed far from the trade he had learned in his youth. He worried about detail and small economies, proceeding methodically step-by-step, employing the latest technology available and building a small concern into one of the larger houses in its field. These are the qualities of a good 19th century business man; conservative, attentive, steady, working with what is familiar and improving on it. While present in any good businessman of the age, these characteristics were, however, not the signal qualities of Horstmann's two sons.

William J. and Sigmund H. Horstmann now assumed complete management of the business. Under their direction the business had already become highly dynamic. While also building on the traditional business of manufacturing woven lace, during the next 25 years under their direct management, the Horstmann Company was thoroughly reorganized. A new state-of-the-art factory was built, and in that factory new types of manufactures supplemented the firm's traditional manufacture of lace and trimmings. In that quarter century the Horstmann firm became the largest manufacturer and dealer in military goods in the United States.

[82] *One Hundred Years,* p. 27.

Fortunately, we now also can rely on actual records of the firm. Its machine book covers the tools from 1842 to 1870. Sample books of product labels exist for the period 1848 to 1875. A cash book exists for the Civil War era, as do the actual balance sheets in manuscript for both the Philadelphia firm and its New York branch house. Labor records and partnership agreements allow us to finally be on the inside looking out rather than the outside looking in.

The first important change made by the brothers dates to 1843, even before Horstmann's retirement, with the creation of the new firm of William H. Horstmann & Sons. An inventory of the tools of this firm taken two years later on Horstmann's retirement February 1, 1845, shows the value of its machinery to be $2,971.54. The brothers added only $725.80 to this capitalization in 1846. However, in 1847 their investment in tools was $4,822.48, more than doubling the firm's capabilities. From then on, until 1860, the value of the machinery added was never less than $2,496 (1850), nor more than $5,237 (1855), totaling by 1860 a value in machinery alone of $75,563.48 including a single expenditure of $25,000 to purchase the Clinton Company, a large textile manufacturer in Clinton, Massachusetts.[83]

Also included within this capital investment were the purchases of the tools of Frederick W. Widmann and John O'Harra. While the Widmann purchase was only $200, it included the tools and machinery of his sword shop, a type of manufacture which Horstmann senior would almost certainly never have added on his own.

A fellow German, F. W. Widmann had arrived in Philadelphia on October 23, 1816, aboard the Brig *Hannah* from Bremen with one trunk, one knapsack, and four boxes of tools.[84] From 1816-1848, he had succeeded in Philadelphia in a small way as a sword mounter. Significantly, Widmann never made sword blades, but simply imported them from Germany and mounted them in gaudy brass hilts and scabbards. These included rolled bands of decoration in pressed brass with leaf motives which were used to decorate both sword hilts and scabbards. William H. Horstmann & Sons continued Widmann's tradition of importing blades and making mountings in their own "sword shop."[85]

Not only did the Horstmann brothers purchase Widmann's tools, they

---

[83] Coles Lippincott Horstmann Papers. Machine Book, p. 1.

[84] National Archives: *Lists of Vessels Arriving at Philadelphia 1800-82*, Microfilm number 425, Roll 23, List 262.

[85] Bazelon, Bruce S., *Swords from Public Collections in the Commonwealth of Pennsylvania* (Lincoln, R. I.: Man at Arms, 1985). Chapter on F. W. Widmann.

also hired a number of his workmen. Unlike the concise wills made by the Horstmanns, Frederick Widmann's will, dated September 10, 1846,[86] is that of a small businessman quite concerned with detail. He names design books, and armorial drawings "... also the book on electro-galvanic gilding ..." to go to Louis Loemmel conditional on the latter staying with him. Widmann goes on to bequeath winter and summer clothing to Julius Knocke, "if he remains at my house". Although Loemmel did not actually receive the Widmann's design books because he had left Widmann's employ by March of 1848–a month before Widmann's death on April 12, 1848–the will does show that Knocke and Loemmel were regarded as skilled workmen. Horstmann's 1868 employee register lists Knocke and Louis "Lemmoel" as having been hired May 15 and May 8, 1848 (respectively), within one month of Widmann's death. Lemmoel had previously left Widmann's employ but this cannot be tied to working for Horstmann previous to May, 1848. An Albert Lammoel was hired in Horstmann's sword room on October 10, 1850, presumably a relation of Louis Lammoel.

A Charles Klobe was hired by Horstmann in September, 1849, but he has no known relationship to Widmann. Widmann's design books went to an apprentice, Jacob Faser, who may have worked for Horstmann for a few years and then moved south.[87] Faser was a German who had immigrated in 1828 and who, after his Philadelphia stay, settled in Macon, Mississippi. In 1861 Faser worked for James Conning in Mobile, Alabama, where he supervised Conning's sword shop by at least June, 1861. However, in 1862 Faser had left Conning and was back in Macon where he was later elected mayor.

One other Horstmann note appears in Widmann's will. He mentions in the codicil of March 22, 1848, which disinherits Lammoel, "... Mr. Horstmann can have back again the silver I got of him at the same price I paid for it. It lays uninjured (sic) as I received it of him in my store in the sideboard there." All this speaks to some sort of ongoing business relationship between Horstmann and Widmann. Although clearly Widmann hoped that his workmen would keep the business going after his death, he provided that it be converted into ready money if this was not to happen. Clearly the Horstmanns purchased the Widmann busi-

---

[86] Philadelphia Will Book 20, No. 97, p. 203. Philadelphia City Archives.

[87] The American Society of Arms Collectors *Bulletin Number Thirty-Eight*, Spring 1978, "James Conning, Confederate Sword Maker" by R. E. Neville, Jr., pp.13-20. Also, conversation with Clarence Runtsch about Faser's subsequent career.

ness when the business could not be sold internally. Widmann's will was written in German and includes a note from the translator certifying its accuracy. When the estate was settled the remainder went to Widmann's sister in Breslau and then to his brother's children. The administrator, a nephew, lived in Bremen, the same port from which Widmann sailed in 1816.[88]

A drawing book depicting swords made by F. W. Widmann is in the collection of Mr. Norm Flayderman. This book shows swords from 1846 through July, 1848, several months after Widmann's death in April, 1848, and after at least two of Widmann's chief workmen were hired by Horstmann in May. This book contains what would be called today work drawings of a number of swords and most probably accompanied Widmann's tools and workers into Horstmann's business.

The "boys" also bought "fall press, dyes &c. for U. S. Cap Mountings &c" from John O'Harra for $100. O'Harra enjoyed federal contracts for insignia in the late 1830s and 1840s. After the initial transactions in 1848, the Horstmanns began to add new dies in order to increase their manufacturing capabilities in swords and stamped metal insignia.[89] Many of these insignia dies still exist. The Horstmann company later employed the firm of Thomas Parry & Sons to do its stamping. Though the dates of Parry's business with Horstmann are unknown, they must date after the firm's inception, listed in the Philadelphia city directories as 1874.

When Parry's tools were sold at auction in the early 1920s, Horstmann's dies went to Hahn companies of Baltimore and to W. Stokes Kirk & Company of Philadelphia. By that time the insignia dies also included those purchased from the estate of William Pinchin & Company of Philadelphia, the heir to George Armitage's stamping business, whose firm was purchased by Horstmann on Pinchin's death in 1862.[90]

Horstmann's new manufactures also included drum making and flags. The machine book notes in June of 1850 the purchase of "two planes for drum shop," our first notice that Horstmann was actually manufacturing drums. The flag shop was also in operation before the Civil War. The surviving color of the Logan Guards of Lewistown, Pennsylvania, was carried to the first action of the war in April, 1861, when

---

[88] *Ibid.*

[89] Machine Book entry January, 1849 listing accounts brought over from 1848 Machine Book entry January, 1849, listing accounts brought over from 1848.

[90] Machine Book entry August 1862 "Lot of machinery and tools belonging to estate of William Pinchin obt. At sale of M. Thomas & Sons, $139.52."

this unit marched to Washington, D. C. to guard the nation's capitol. The Logans were formed in 1859, and their flag, which is signed Horstmann, probably dates to that time.

In the meanwhile, the Horstmanns continued to consolidate their business. The establishment at 51 North Third had been extensively rebuilt and expanded, but was immediately too small to contain Horstmann's burgeoning business. A drawing of this structure appears in R. A. Smith's *Philadelphia As It Is in 1852.*[91] This drawing shows a five story building with a sales room for military goods on the street floor and a sign two stories above, "Manufactory of Laces, Fringes, Bindings, Cords, Tassels &c". The accompanying advertisement shows Horstmann selling about everything in the military line including pistols but with no mention of uniforms.

William J. Horstmann's ladies trimming store was at 204 Chestnut from 1851 to 1853, and in 1854 moved to Seventh and Chestnut. In 1855, he was at 223 Chestnut. This last address became the main sales operation of the store in that year, and the 51 North Third Street location was apparently sold after the final move to 223 Chestnut on July 1, 1856.

The biggest Horstmann enterprise, the building of an entire new factory complex at Fifth and Cherry, was considered remarkable in 1853 when the factory opened. It was also an unusual move for its day. Philip Scranton makes the point that Horstmann is unique of the large textile manufacturers in moving the main factory complex into Philadelphia from Kensington rather than moving out to one of Philadelphia's outlying communities.[92] For textile manufacturers these communities were Kensington, where Horstmann already had a factory, and Roxborough. The building at Fifth and Cherry is described in the *Daily Register* of October 22, 1853, as follows: "It has a front of 139 feet and 50 feet depth on Fifth Street and a 98 foot front and 40 in depth on Cherry Street, having a space in the rear for an engine-house and machine shops. This building is 75 feet long and 25 wide, two stories high, while the main building is five stories with a basement . . . there are two stairways of ample dimensions . . . about 1,300,000 bricks were used in the construction of the building . . . the engine is capable of being worked to 160 horsepower."[93] The *Philadelphia Sunday Ledger* article of July 3, 1853, describes the building as:

[91] Smith, R. A.: *Philadelphia As It Is in 1852* (Phila.: Lindsay & Blakiston,1852), p. 376.

[92] Scranton, Philip: *Proprietary Capitalism: the Textile Manufacture at Philadelphia, 1800-1885* (Cambridge: Cambridge University Press, 1983).

[93] Coles Lippincott Horstmann papers. Horstmann scrapbook.

". . . a chaste, stupendous brick structure occupying 100 feet on Cherry Street and 140 feet on Fifth Street, of six stories high. Daylight is let into this commercial palace through 180 windows. . . . Inside is an exciting scene of orderly bustle; 300 handsome girls and 200 good looking workers of the sterner sex are hard at work in all the little delights of enthusiastic shoppers. The great bulk of "all this sort of thing" - (the ladies understand the intricate details of trimmings much better than we do) - sold in the United States emanate from this human hive.

"In conjunction with these gentle productions, is a savage department. However, Mars and Ma's is an old association. This savage department is devoted to the manufacture of all sorts of military and naval trimmings. There is one blood-thirsty looking room chock full of swords - these are imported; the scabbards and decorations are made in an adjoining room. There is also a Drum department - about two-thirds of the drums used in the country sound their reveille in this establishment.

"A powerful steam engine sets all the machinery in motion. . . . There is no fire on the premises except that in the engine house, which occupies a detached building; the rooms are warmed by means of hot vapor pipes. . . . Mr. Hoxie, the architect has provided the most complete means of human egress. A communication is to be established to all parts of the immense building by means of speaking tubes, - and packages be sent from the cellar to the sixth story, by means of machinery. The drainage is arranged on the most scientific principles; in fact take it all together, this is just such a place as old cholera must regard with a most malignant eye - how to baffle the well regulated ventilation and cleanliness must be a puzzler. . . ." [94]

While buildings such as the Horstmann's commercial palace, as described in 1853, are quite commonplace to us, having seen the hundreds of mills left over from the early 20th century, in fact Horstmann's new building at Fifth and Cherry was a marvel in its day, and the precursor of the many factories which would follow. It is so well remarked upon in the Philadelphia papers because in fact for its time it was unique, no matter how commonplace such buildings would later become. The building was designed by J. C. Hoxie with the active participation of William J. Horstmann in design and construction.

[94] Coles Lippincott Horstmann Papers. Horstmann scrapbook.

One early innovation to the new factory was the addition of power loom machinery. This was obtained through the purchase of the Clinton Company of Clinton, Mass., in January, 1857. A broadside inserted into the Horstmann scrapbook bearing this date states that Horstmann had purchased "... all the Power Loom Machinery, Patent Rights and interests of the said Company connected with the manufacture of Coach Laces by Power Looms. The looms are now in the course of removal to our Manufactory at Fifth and Cherry Streets and we shall hereafter continue the business in this city". The move was seen as an important one. An article in the scrapbook taken from "The Hub" of January 1, 1877, while seen from the perspective of twenty years later, shows the introduction of the power loom as a remarkable technological innovation.

All of Horstmann's various manufactures, the traditional lace and trimmings business, the new manufacture of insignia, drums and flags as well as the hilting of sword blades were now brought together at Fifth and Cherry, as the old Columbia factory in Kensington was sold in 1853, the same year the new establishment opened. Only the salesroom now remained outside this complex. In 1858 the salesroom moved to 723 Chestnut. The sales operations were still at 723 Chestnut in 1862 but they too moved to Fifth and Cherry as soon as additional land was secured. Thus, the entire firm was at the new location by 1862.

The *Hexamer General Surveys* contain insurance drawings of the Horstmann factory at various times beginning in 1865 and ending in 1916. These show a complex of rooms belonging to Horstmann Brothers & Company and William H. Horstmann & Sons. The 1865 survey lists the carding, callendering, and ware rooms in the basement; the sword shop, office , tassel, and other departments on the first floor; the ribbon and "ritteling" departments on the second floor; the coach lace department and silk room on the third floor; the braid ribbon and warping departments on the fourth, and cording etc. rooms in the attic. These functional descriptions of the building were simply copied verbatim in the surveys of 1866, 1872 and 1887. The survey of 1872 noted that the wood shop produced very few shavings and employed only two hands. In that year the factory employed 300 hands and the salesrooms another 75. Of these employees 125 were listed as male and 250 female.[95] The 1887 survey notes that the factory employed 500 hands and an additional 100 worked in the sales rooms. The work force is described as one-fourth male and three-fourths female. The survey also notes that

[95] *Hexamer General Survey* October, 1872, pp. 626-627. E. Hexamer & Son, Philadelphia. Philadelphia Free Library. Map Room.

the factory produced silk and worsted goods utilizing worsted yarn and silk thread. No yarn was manufactured at this time though, as we shall see, by the early twentieth century it would become the main product of the company.[96]

A letter dated October 6, 1855, from William H. Horstmann & Sons to the Waterbury Button Company survives in the files of the Mattatuck Museum's Waterbury Companies files. One page of this letter is a broadside depicting the new factory building and giving a listing of the goods for sale and clearly shows the transition from 51 N. 3rd to the new factory.

> "William H. Horstmann & Sons, Manufacturers & Importers of Military Goods, No. 51 North Third Street, Philadelphia. Manufactory, Corner of Fifth & Cherry Streets. A complete and elegant assortment of the following articles constantly on hand: swords, belts, sashes, caps, plumes, epaulettes, drums, pompoms (sic), gloves, spurs, cloths, holsters, bridles, buttons, saddle cloths, sword knots, cassimeres, cartridge boxes, bayonet scabbards. Banners and flags for presentation. Masons, Odd Fellows & Sons of Temperance Regalia. Regalia Makers furnished on the most advantageous terms with Trimmings and Embroideries. Gilt carved eagles, gilt and plaited braid, openwork laces, silk and bunting flags, gilt and plated bouillon, silk and cotton velvet, gilt and plated gimps, banner and flag silk, spangles, tassels, bouillon fringe, spangled and embroidered stars, gold and silver leather, banner poles and belts."

This letter shows a clear definition between what was sold as merchandise and the items being manufactured. One significant feature is that uniforms are not mentioned in either category and it would seem that the Horstmanns put off dealing in actual tailored garments as long as possible while a large part of the business consisted of trimmings for both uniforms and clothing.

With the new factory came reorganization of the business. From 1828 Horstmann had really been conducting two businesses, trimmings and military goods. An organizational structure was created in the 1850s which reflected this duality. William H. Horstmann & Sons had been created in 1843 out of the initial business. Now, a new firm, Horstmann

[96] *Hexamer General Survey* October, 1887, pp. 2093-2094. E. Hexamer & Son, Philadelphia.

Brothers & Company, took on the military side of the firm, leaving William H. Horstmann & Sons to carry on the trimmings trade. The new firm is first listed in Boyd's Co-Partnership Directory in 1858. This directory names the partners as William J. Horstmann, Sigmund H. Horstmann, H. V. Allien, and John G. Franklin. The 1859 Philadelphia City Directories show the new firm, Horstmann Brothers & Co., had not changed from 1858 while the older firm, William H. Horstmann & Sons, now consisted of William J. and Sigmund Horstmann. A third firm, William J. Horstmann & Co., included the two brothers as well as George O. Evans and William S. Hassall. The Horstmann Brothers firm and William J. Horstmann's trimming business were located at 723 Chestnut while the older firm, William H. Horstmann & Sons, had already relocated at 5th & Cherry. The R. G. Dun and Company rating of the time gives an interesting perspective to this business decision.[97] The entry for January 1, 1856 states ". . . will remove to Chestnut St, and give up the retail store" while the December 23, 1856 entry notes that they are ". . . a wholesale bus. only. . . ."

It may be that the Horstmanns were thinking of simply manufacturing and wholesale supply when their factory was first built. If so, the notion did not last long and salesrooms were soon added to the new factory.

The new members of the firms–George O. Evans, W. S. Hassall, John G. Franklin, and Henry V. Allien–reflect the growth of the business beyond the simple partnership of a father and his two sons. It was the usual business custom of the period that younger individuals wanting to enter a trade or business served an apprenticeship. While most of the formal aspects of this apprenticeship process had broken down by the mid-19th century many of the residual practices remained. Management employees, while never formally apprenticed, were expected to have "gotten their feet wet" by working in an already up and running business. R. G. Dun and Company's highest compliment in assessing business ratings was to refer to someone as "brought up in the business."

The surviving draft of the partnership agreement between the Horstmann Brothers, Franklin, Evans, and Hassall, now in the Lippincott Papers at the Historical Society of Pennsylvania, clearly shows that the junior partners brought no capital into the business but were entitled to 5% of the net profits, not less than $1666.66 each. As an added incentive, and phrased exactly in those terms, if any of the junior partners

[97] R. G. Dun and Company Collection. Phila. 1.511. Entries January 1, 1856, December 23, 1856.

were to die, the $1666.66 would be paid into his estate for the full calender year. R. G. Dun and Company notes on January 1, 1855, "this day admitted Wm J. Hassall." The entry for January 30 postscripted that "Hassall added no capl" (capital), a most important partnership consideration.[98]

Henry V. Allien (1814-1889) managed the firm's New York branch after the dissolution of a partnership between the three Horstmanns and Gustavus Drucker, a New York importer, in 1850. Drucker was first listed at the firm's address of 36 Platt in 1836 and is listed in the New York city directories as a partner from 1845 to 1850. However, the separate co-partnership directories for New York show Drucker was already a partner in 1843. These listings reflect virtually Drucker's entire business history in New York. When the firm was dissolved on February 1, 1850, Henry V. Allien was already one of the persons named who would settle the affairs of the firm.

What is known of Allien is contained largely in his obituaries in the New York papers August 15, 1889, which appear in the Horstmann scrap books. The Centennial History refers to him as "... a gentleman of French descent related to an old friend of Mr. Horstmann living in the commercial capital. ..."[99] "He was the son of an officer in the French Navy who, while stationed at San Domingo about the end of the last century, married the daughter of a planter, and after the uprising of the blacks there, moved to this city. Henry V. Allien was born here in 1814. He began his commercial career as an employee of the old firm of Brugiere & Alien, silk importers ... the junior member of which was his elder brother. Afterward he was with Jacquelin & Allien, in the same line, and eventually he became the New-York representative of William H Horstmann & Co of Philadelphia, the pioneer silk manufacturer of this country. ..." The partnership would last 27 years.[100] H. V. Allien would assume full financial control of Horstmanns' New York operations in January, 1877. This move was clearly an amicable separation showing the Horstmann family's evident desire to retrench following the death of the Horstmann brothers.

During the early days of the Civil War, George O. Evans and William S. Hassall split from the Horstmanns to found their own firm,

---

[98] R. G. Dun and Company Collection. Philadelphia 1.511. Entries January 1, January 30, 1855.

[99] *Centennial History,* p. 16.

[100] *New York Times* and *New York Daily Tribune* obituaries in Horstmann scrapbooks. Both dated August 15, 1889.

Evans & Hassall. This firm lasted through the war years and later evolved into J. H. Wilson & Co, a competitor of the Horstmanns' through the rest of the century.

There is a difference between the early Horstmann partnerships and those in force after the Civil War. Excepting Franklin, who stayed with the business, George O. Evans, William S. Hassall, and Gustavus Drucker, having apparently put in little initial capital, left the business with the funds they made during their association with Horstmann. The details of Henry Duhring's probable partnership are unknown, though he clearly made money during the years in question. Evans and Hassall, having gotten their trade educations, left to found their own firm. Drucker simply fades from the picture when his usefulness ceased. However, Horstmann's later partners–Allien, Clarkson, and Weihenmayer–were an integral part of the Horstmann business and wound up well off after having stayed with the business many years. This may reflect the changing nature of business in the United States, where partnerships were no longer made and broken with such great regularity as they were before 1860.

Men like John G. Franklin, Samuel Clarkson (cited as William J. Horstmann's brother-in-law in Horstmann's will), J. Howard Mecke, and Albert Weihenmayer are nonentities to us because they never ran their own firms, yet they represented the senior management of Horstmann during the third quarter of the 19th century. They left the trade far wealthier than many whose names are much better known.

John G. Franklin joined the Horstmann firm about 1843 and stayed with them for more than forty years. Weihenmayer came about 1853 and stayed until his death in 1887. Mecke, a Union Army officer, came to the firm after the Civil War and stayed until his own death in October, 1887. Along with Samuel Clarkson, these men represented the team which Sigmund and William J. Horstmann assembled to help manage their company. The Horstmann business, while incorporated in 1894, remained essentially a family partnership until 1923. However, the firm was far more complex than in the 1840s, when William J. Horstmann built his own looms while brother Sigmund kept track of the firm's finances and stock under the benevolent eagle eye of their father.

The situation of Horstmanns' other employees is less clear. By the Civil War there were hundreds of them. An employee register begun in late 1867 and actively maintained through early 1875 cites the dates of employment of all members of the firm. While one employee is listed in 1867 as having started in 1828, and others were employed since the 1830s, the majority were of much shorter duration, most lasting a few years or, in some cases, a few months. What are described as "contract numbers" given to each employee from at least as early as the late 1840s run through 3650, while approximately 1750 of these are listed as active sometime during the register's dates of use (1867-75).

The following table depicts the various trades in the Horstmann factory complex at Fifth and Cherry. The employees' time book kept by the firm seems to show work functions relating to both William H. Horstmann and Sons and Horstmann Brothers and Company. Those employees still with the firm in 1867 were also entered in this ledger along with their contract number, date of employment, portion of the factory they were assigned to, their date of leaving, and the reason why. The following table compares the functions of the firm in 1868, the first complete year given in the ledger, with those in 1872, the median point in the record, showing the number of employees assigned to each area.

TABLE 4
Distribution of Occupational Specializations, 1868 and 1872
Horstmann Factory, Fifth and Cherry Streets

Factory Area	Workers, 1868	Workers, 1872
At home	3	11
Attic cord walk	17	25
Basement cord (walk)	3	1
Basement	0	7
Basement flag department	4	4
Basement shade tassel department	15	3
Basement Prahls [named for s.t. supervisor]	4	5
Basement tassel room	0	3
Bast twister	0	1
Braid machines	1	0
Braid machines	1	0
Braid room	14	36
Burnishing	0	2
Coach lace room	12	9
Counting room	1	1
Drum shop	1	1
Engineer	0	1
Fourth Storey spool	0	1
Hand looms	47	39
Loom shop	1	1

Machine shops	9	3
Machinist	0	4
Maurer's room (hand looms)	0	1
Office	16	18
Quill winding for power looms	1	18
Packing wool	1	1
Porter	0	1
Power loom quill frame	1	1
Power looms	60	59
Power loom room	0	6
Quill winding	21	11
Raw silk winding	1	0
Reeler	0	1
Sewing room	34	51
Shade tassel department	112	16
Silk room	21	30
Spooling of belt	0	1
Spooling frames	11	14
Spooling for braid machines	12	14
Spooling room	0	7
Stringing beads	1	2
Store room flag	2	4
Store room	7	3
Sword department/room	6	6
Sword shop	7	11
Taking down coach lace looms	1	0
Tassel room	44	131
Twisting wool	0	6
Ware room	3	0
Warping room	12	17
Watchman	1	1
Werner's room (Joseph Werner, sewing room)	0	1

It is evident from the above table that the military artifacts produced by the firm represented only a small part of Horstmanns' overall product. This assumes that the labor force is an accurate indicator of the actual product of the Horstmann factory. The sword shop, the drum shop, and the basement flag room together do not have anywhere near as many people as were occupied in making shade tassels, coach lace, or working the looms, making essentially civilian products.

The number of employees of the Horstmann firm cannot be ascertained until the 1850s when newspaper accounts begin to describe the firm as having a large number of workers. The *Sunday Ledger* of July 3, 1853, cited above, notes that there were 300 girls and 200 men at that time. However, the *Philadelphia Daily Register* of October 22, 1853, states that the Horstmanns employed 250 female and 100 male workers, a large discrepancy. The only indication from the Horstmann employment book is that the first contract number noted, in 1848, is for number 204, which would tend to give some credence to the *Daily Register* account. The *Register* continues in stating that the girls received between two and six dollars weekly, the boys between two and four dollars weekly, and the men from six to fifteen dollars.

The following table represents the employees who can be identified in the late 1860s through the mid-1870s by their date of employment:

TABLE 5.
Employee Retention, 1867 - 1875
Calculated from Original Year of Employment

Year of Employment	No. Workers	Year of Employment	No. Workers
1828	1	1858	5
1833	1	1859	8
1836	3	1860	9
1837	1	1861	10
1838	1	1862	5
1839	2	1863	21
1846	1	1864	21
1847	2	1865	32
1848	5	1866	39
1849	2	1867	178
1850	7	1868	224
1851	5	1869	182
1852	1	1870	99
1853	6	1871	194
1854	5	1872	275
1855	12	1873	194
1856	18	1874	113
1857	10	1875 (partial)	71

The table on the previous page emphasizes the essential stability of Horst-manns' core work force. It is equally evident that this force was easily supplemented by short term employees as business conditions demanded. The employees hired in years previous to 1867 are only those personnel remaining from a presumably larger work force. The register apparently was begun in the second half of 1867, and that year includes all employees hired prior to that time who were still at the Horstmann factory in the second half of that year. Short-term hires are invisible in these data, except for the years 1868 to 1874, 1875 also being a partial year. The figures for 1868 through 1874 show the number of people who were actually hired in those years.

Table 6 shows the employees who can be identified as having worked in the military industries run by the Horstmann firm and their employment records. Here again the senior employees appear to have worked as foremen running the day-to-day operations. In the sword shop Julius Knocke and Louis Lammoel started with F. W. Widmann in the 1840s, joining the Horstmann business when the Widmann firm was purchased in 1848. They were still with Horstmann in the 1870s.

TABLE 6.
Profile of Work Force between 1867 and 1875,
Sword Shop, Flag Shop, Drum Shop

Worker	Work Location	Date Hired	Date Left	Reason for Leaving
Domer, Gustav	sword shop	12/18/72	9/27/73	left w/o notice, going to NY
Barr, Wilhelm	sword dept.	4/6/70	7/19/73	sent away, could not account for some silver, was told there was no work
Clemann, Simon	sword shop	5/28/70	4/26/72	gave notice, quarreled w/ Mr. Knocke
Erb, Gottlieb	sword shop	5/21/60	1875+	
Echter, Julius	sword shop	6/16/71	8/22/71	left, wanted 11.00 per week. Did not think him worth it.
Ewald, George	sword shop	6/20/73	7/18/73	Out of work.
Frey, Lena	sword shop	3/24/73	12/4/74	Out of work.
Goltermann, Frederick	sword shop	4/24/60	4/11/70	Gave notice to leave.
Hiss, John Paul	sword shop	9/13/56	3/21/73	Discharged. Mr. Schultz thought he did not ----
Herbstrith, Hermann	sword shop	8/1/70	2/2/71	Out of work.
Brand, Christian	sword shop	3/1/72	12/2/72	Sent away + Walter Dougherty took his place.

Dougherty, Walter	sword dept.	9/70	3/2/72	Going to learn a trade-wheelwright
Dougherty, Walter	sword dept.	ret. 1/2/73	5/31/73	Sent away.
Beste, Eugene	sword dept.	6/11/72	4/12/73	Gave notice. Poor hand.
Beste, Eugene	sword shop	6/8/74	1875+	
Johnson, William H.	sword room	12/63	1875+	
Jackel, William	sword shop	6/69	3/2/72	Going to keep store.
Fouser, William	sword shop	6/69	6/7/70	No reason given.
Billy, Harry	sword shop	5/27/74	9/11/74	Gone to charging room, store. Resigned and discharged 5/75.
Baudoy, Otto	sword shop	6/13/74	8/7/74	Out of work.
Grettenberger, Fred. K.	sword shop	2/18/73	5/10/73	Dissatisfied with piece work.
Gorlich, Ernst	sword shop	1/6/74	7/30/74	Out of work.
Knocke, Julius	sword room	5/15/48	1875+	
Klobe, Charles	sword room	9/1/49	1875+	
Kirchner, Johannes	sword dept.	5/20/71	4/27/73	Gave notice. Dissatisfied w/ salary.
Lammoel, Albert	sword room	10/3/50	6/16/71	Gave him notice to quit + discharged w/out pay, took tools not belonging to him.
Lammoel, Louis	sword room	5/8/48	10/21/67	Died in our blade magazine at 2/1/20. Went to lie down + closed the door, saw through the window that he was dying.
Lautenbach, Frederick S	sword shop	5/12/70	7/29/70	Gave notice to leave. Not a good hand.
Lange, Harry	sword shop	7/11/72	10/4/72	Mr. Knocke could not get along with him.
Lambach, Harry	sword shop	4/28/73	1875+	
Lerch, Mauritz	sword shop	6/69	6/23/69	Left.
Muller, Joseph	sword dept	9/6/70	11/18/70	Was told to look for another place.
Muller, John	sword dept	10/19/71	12/16/71	Gave notice to leave.
Mertz, Rudolph	sword shop	9/12/74	1875+	
Martenssen, S	sword shop	5/16/70	12/10/70	Out of work, not steady nor sharp, engr. scabbards, but slowly.

O'Brien, John	sword room	3/18/50	11/70	Out of work.
Michel, Frederick	sword shop	9/29/73	5/16/74	Gave notice.
Ihrig, Jacob	sword shop	6/69	6/19/69	Left.
Plantholtz, Adolph	sword room	9/1/51	1875+	
Pohle, Theodor	sword dept.	8/67	1/21/68	Gave notice to leave, goes back to his home in Rhode Island, his property a large hotel being restored to him.
Ramke, Henry	sword shop	7/68	6/17/72	Left without notice.
Ramke, Henry	sword shop	5/20/73	7/14/74	Out of work.
Ramke, Henry	sword shop	ret6/22/74	8/7/74	Out of work.
Roth, George	sword shop	1/19/71	5/7/72	Going to leave the city.
Schneider, Charles	sword dept.	7/70	10/1/70	Son of Ramke. Left to go to his uncle in the country.
Schultz, Louis	sword shop	4/3/68	5/18/68	Started a shooting gallery at Washington's Retreat.
Schorr, Constantin	sword shop	6/68	2/13/75	Out of work. Found work elsewhere.
Turner, Maggie	burnishing, sword shop	7/15/70	1/2/72	Left on a/c of her health.
Todd, Jane	sword dept.	3/24/73	8/20/75	Gave notice.
Schultz, Louis	sword shop	10/12/72	7/2/75	Left on acc. of bad health, gone to Niagara Falls.
Schwechler, John	sword shop	5/27/73	3/27/74	Gave notice. Going to learn a trade.
Walther, August	sword room	9/16/70	1875+	
Wolter, Paul	sword shop	7/72	1/24/73	Sent him away. Paid no attention to his work + too much to the girls.
Wenheim, Heinrich	sword dept.	4/25/70	1875+	
Collom, Albert	drum shop	7/47	11/2/72	Reason unknown.
Collom, Albert	drum shop	???	2/17/74	Goes into business w/ Mr. Long. Machine repair.
Bishop, Sallie	bsmt flag rm	1869	5/22/74	Left. Believe she got married.
Dudley, Martha	store room flag-dept	1856	6/6/68	Discharged for misbehavior.
Dudley, Martha	store room flag-dept	???	5/9/70	Gave notice: got married.

Hofman, Anna M.	bsmt flag dep	6/69	1875+	
Murray, Ella	bsmt flag dep	6/12/67	1875+	
Clark, Annie	store room flag-dept	12/8/65	1875+	
Annely, Margaretta	bsmt flag dep	5/24/75	1875+	
Barrington, Carrie	bsmt flag dep	5/24/75	1875+	
Walters, Nellie	bsmt flags	6/24/67	1875+	Gave her work in sewing room.
Eckert, Caroline	store room flag-dept	1850	6/20/68	Left on account of ill health.
Julian, Alice	bsmt flag dep	9/2/68	5/6/69	Gave notice to leave.

In the 1851 Philadelphia city directory Knocke appears as a die sinker and Chaser at 8 Sansom while Albert Cullom was a drum manufacturer, John O'Brien a brass worker, and Louis "Lammel" was a chaser. As we have seen, Knocke and Lammoel had worked for Frederick W. Widmann before coming to Horstmann, presumably shortly after Horstmann purchased Widmann's tools, while Cullom and O'Brien appear to have had similar independent trade skills even though their long term employment was with Horstmann.

From the foregoing table it is evident that Horstmann employed a few skilled workmen and picked up other hands on piece work. Many of these were dissatisfied, and their careers with the firm were short. Given their last names, the overwhelming majority of workers in the sword shop seem to have been of German extraction. We can also infer that of the businesses which the Horstmann brothers added to the firm in the late 1840s, only the sword shop had grown and seems to have been prosperous through the late 1860s and 1870s. No mention is made in the employee register of producing insignia, just as none of the dies necessary for insignia manufacture appear in the machine register after 1861. The drum shop, which is referred to in the *Public Ledger* account of 1853 as being a substantial operation, by 1867 is maintained by one employee.

The flag department, while prosperous during the Civil War, may have retained its employees but work was still carried on by only a handful of people. In all probability Horstmann had always contracted a portion of this work outside his factory. Certainly it proved cheaper in the long run for Horstmann to do what it had always done, namely subcontract pieces requiring specialized manufacturing capability, or to simply buy them from other manufacturers. The cash book of Horstmann Brothers & Co. (1858-60), which will be examined in more detail below, lists the names of many other military manufacturers with which Horstmann was doing business, mostly purchasing merchandise to be resold. Among these are the Lane Mfg. Co. of Waterbury (buttons) , William Pinchin (insignia), William H. Gray (insignia), Charles Oakford (hats), and William G. Mintzer (probably trimmings).

The Horstmanns had now consolidated their business, including salesrooms, at Fifth and Cherry, giving up the salesroom at 723 Chestnut in 1861. The venerable 51 N 3rd location was also gone. The 1865 McElroy's *Philadelphia City Directory* contains a page advertisement citing the two firms.[101] William H. Horstmann & Sons were "Manufacturers and Importers of Ladies' Dress, Cloak and Mantilla Trimmings. . . ." Horstmann Bros. & Co., are listed as "Manufacturers and Importers of Military Goods of all kinds. Masonic Regalia, Flags, Banners, Drums, &c." Both continued to show their addresses and salesrooms at Fifth and Cherry. The Hexamer Insurance Company maps at the Philadelphia Free Library contain two plans of the Horstmann company dating about 1865 and 1866. Both show the civilian and military manufacturing rooms of William H. Horstmann & Sons and Horstmann Brothers & Company intermingled but in distinct clusters within the factory.

One clear way of understanding what the Horstmanns produced in the 1850s and 1860s is through their label sample book, another of the manuscript papers and company scrapbooks held by the Historical Society of Pennsylvania in its Horstmann-Lippincott files. This book shows samples of labels originally fixed to the packaging of items or to the actual specimens themselves (such as drums). It depicts each label designed from 1848 to the mid 1870s, when ordered, and how many. The labels include military items interspersed with the many types of tassels, ribbons, and trimmings manufactured by Horstmann's civilian business. Many of the latter label types are for the Columbia Manufacturing Company, a trade name the Horstmanns continued to use though the Columbia factory had been sold.

In May, 1851, the first military tassels are shown in the label book. As the new U. S. Army Regulations of June, 1851 and the U. S. Naval Regulations of March, 1852 changed the appearance and the arms of American soldiers and sailors, the Horstmanns were prominent in publicizing the new styles and in quickly switching over to new patterns. This merchandise is amply shown in the label book. A little later (1857)

---

[101] McElroy, A: *McElroy's Philadelphia City Directory for 1865* (Philadelphia: A. McElroy,1865). Advertising section, p. 5.

Horstmann issued a color lithographic book showing the newly adopted uniforms. This book was in no way an official publication. A letter dated November 22, 1857, was drafted by the Horstmanns presenting Colonel Craig of the U. S. Army Ordnance Department with a complimentary copy of this new book of Army regulations.[102] Both the increased military trade shown in the label book and the active solicitation of military customers evidenced in the fine uniform plates of Horstmann's 1857 regulations book document the aggressive pursuit of an increased military trade during the 1850s which resulted in the creation of Horstmann Brothers & Company in 1858.

Sword labels first appear in the label book on May 8, 1852, and drum labels in December, 1853. Other labels printed in these years include those intended to go on boxes containing worsted non-commissioned officer, corporal, and privates cloth epaulettes (June 5, 1852), U. S. pompons (June 5, 1852), military silk and worsted sashes (February 23, 1852), military cap eagles (September, 1854), military belt plates (April, 1854), and such items as military plume sockets, shoulder scales, cross cannons, and belt plates (November, 1854). Flags, tassels, cords, and other items of muslin and silk are prominently cited both in November, 1854, and March, 1855. More military sashes appear in the label samples in December, 1856, while sword belts are shown in March, 1856. Plume sockets and belt plates are shown in February of that year. In June, 1860, labels were ordered for worsted epaulets, shoulder knots, and worsted U. S. hat cords. As the Civil War opened, Horstmann's business in both flags and flag poles can be documented through their label samples. By this date Horstmann's new industries of drum, flag, insignia, and sword making were well established in the Fifth and Cherry Streets factory.

Horstmann's decision to pursue an increased military trade was one of the timely and accurate business assessments which characterize the company's first sixty years. The Civil War represented a very different experience from any which either the Horstmann firm or the country had yet undergone. More than a million Americans were under arms at one time or another during four years of war. Most of the soldiers and sailors in the Union Army were supplied through government contracts in which the Horstmann firms were primary players. In addition, almost all officers were required to privately purchase their uniforms, arms, and accouterments. These officers flocked to the Horstmann sales

---

[102] National Archives. Record Group 92.

rooms in Philadelphia and to Horstmann Brothers & Allien in New York by the hundreds. Even a firm like Horstmann experienced problems with gearing up for war on this scale. Many of its pre-war customers seem to have been from militia units raised in the South. These were now replaced by a new clientele. New manufacturing capabilities had to be developed to handle this expanded product.

Buttons represent one example of this shift. While the Horstmann firm never made buttons, many carry the Horstmann backmark. During the years just prior to the Civil War, most of Horstmann's button business was apparently subcontracted to the Lane Manufacturing Company and later to the Scovill Mfg. Company, both of Waterbury, Connecticut. The Horstmann 1858-60 cash order books clearly show that Lane was the recipient of many button orders at this time. However, a few years later, Scovill's own order books, which begin in 1864, show Scovill doing a large business with Horstmann in manufacturing the military buttons which Horstmann then sold under his own name. In the custom of the time these buttons could be backmarked to either Horstmann, Scovill or with a neutral "quality" mark, such as the ubiquitous "Extra Quality," found on many buttons of the era.

These large button manufacturing companies and a few other large firms had attained a virtual monopoly of the military button business in the United States because of the specialized machinery required for manufacturing military buttons. The large button producing companies, located then in either Waterbury, Connecticut, or Attleboro, Massachusetts, received almost all orders for military buttons in the United States. As expected, the Horstmann machine book records no purchases of dies for buttons, nor any of the specialized machinery involved in button production. At the same time, the records of Scovill now in the Baker Library at Harvard University record a voluminous correspondence with Horstmann from the Civil War period through the end of the 19th century. In May, 1864, shortly after the Scovill order books begin, requests were received from Horstmann Brothers & Allien for staff, Virginia, Maryland, and other buttons. The order of July 8, 1864, from Horstmann Brothers & Co is for 1000 gross of Army coat buttons while their request of November 8, 1864, is for 4000 gross Army coats and 5000 gross Army vest buttons.

Scovill backmark dies, now in the Pennsylvania State Museum, include marks showing various evolutions of the Horstmann firm dating from the late 1860s through the early 20th century. Some obverse dies from the 1880s even have Horstmann's name on the die and the date so that it could be identified as Horstmann property. These dies are depicted in the plates section of this book. Some of the Scovill-made Horstmann backmarked buttons are enlisted-men's types which Horstmann supplied on contract, while many others were intended for officers' uniforms.

The same methodology seems to be true of Horstmann's sword business and insignia manufacture, where again die purchases are recorded up through the beginning of the Civil War, but not after that. It is probable that to supply the increased trade occasioned by the war, Horstmann simply subcontracted much of this manufacture. The low employment figures shown previously for 1867-75 indicate that the sword and insignia shops were small compared to the rest of the Horstmann business. When the trade required, other manufacturers better set up to supply these orders would be contacted. We know, for instance, from surviving specimens, that sword blades with the Horstmann mark, made in compliance with the U. S. Army regulations of 1851, continued to be produced in Solingen, Germany, and then assembled in Philadelphia. It is also probable that woven insignia made in France continued to be imported and sold in the Horstmann Philadelphia sales rooms, along with products of its own domestic manufactory. This particular aspect of the trade is somewhat murky, because while Horstmann did maintain a Paris office, one of his 1852 labels shows a product marked "Fabrique passementier de Paris." These labels could have been intended for attachment to actual imported French goods or to American-made "French goods" denoting their European quality as an advertising device.

Fiscal accounts, probably from papers kept by Sigmund Horstmann, appear in the Lippincott Collection in the Historical Society of Pennsylvania. Within these are balance sheets for the years 1851 to 1863 showing figures for the various Horstmann firms: William H. Horstmann & Sons, Horstmann Brothers and Allien, and Horstmann Brothers and Company. These figures give us a unique glimpse into the actual cash business in which the Horstmanns and H. V. Allien were engaged.

At the same time, these accounts are confusing. In particular, most figures are represented as cumulative totals for the entire history of the corporations. In addition, the gross profit is a fairly small figure, but the shareholders had already taken what they were owed before these final calculations were compiled. Outstanding bills, depending on when they were reckoned, may show some of the operations at a loss for a given year.

The following table shows cumulative cash totals reported by Horstmann Brothers and Allien in New York City, which was sometimes called Horstmann Brothers and Company in its first years, as distinct from the Horstmann Brothers and Company founded in Philadelphia in 1858.

Table 7.
Cumulative Cash Totals
Horstmann Brothers and Allien

Date	Final Cash Total (Cumulative)	Profit and Loss (Cumulative)
1851	$ 99,940.14	$ 22,346.07
1852	$ 90,662.12	$ 36,250.65
1853	$ 94,249.99	$ 56,218.90
1854	$120,334.82	$ 89,606.30
1855	$150,086.44	Figures not available
1856	$172,980.10	$145,737.05
1857	$188,029.87	$186,045.53
1858	$148,305.27	$141,456.15
1859	$159,523.83	$156,395.50
1860	$175,637.11	$170,620.49
1861	$190,008.30	$188,717.40
1862	$246,485.59	$237,183.91
1863	$256,191.32	$242,618.86
1864	$276,523.76	Figures not available

While it is difficult to interpret the seeming losses in the late 1850s, this table does show the effect of the Civil War upon business. The early years of the war record a great deal of activity and then are followed by a rather surprising leveling-off period in the middle part of the war. Regrettably, the figures for the end of the war do not survive.

Table 8 shows figures for Horstmann Brothers & Company of Philadelphia from its inception in 1858 through 1864. This table shows the same uneven progress in the late 1850s together with the tremendous onrush with the beginning of the Civil War.

Table 8.
Cumulative Cash Totals
Horstmann Brothers & Company, Philadelphia

Year	Final Cash Total (Cumulative)	Profit and Loss (Cumulative)	Annual Net Profit
1858	$220,387.84	$ 18,217.47	None given
1859	$ 68,352.33	$ 20,225.84	None given
1860	$ 87,670.20	$ 21,394.76	$ 5,916.25
1861	$111,437.87	$101,162.55	$ 7,512.23
1862	$291,915.36	$ 61,000.62	$51,408.80
1863	$247,618.86	$212,409.72	$31,097.31
1864	$277,192.06	$208,090.38	($1,298.83)

The warehouse accounts for William H. Horstmann & Sons include categories of sales in the company's main line of lace and trimmings. While these may possibly include some military items, military totals cannot be distinguished from the general accounts. The actual sales of military goods in this period were to have been handled by Horstmann Brothers & Company, which had been founded in 1858 to take on the military side of the trade. Because many military items made during the Civil War are marked "William H. Horstmann & Sons" rather than "Horstmann Bros. & Co.", it is evident that the business lines between the two firms became blurred during this period.

Table 9 deals with sales of coach lace, dress trimmings, and foreign imports. It emphasizes that the import trade carried on by Horstmann remained a substantial part of the business even after the creation of the large factory complex at Fifth and Cherry.

Table 9.
Warehouse Accounts by Category of Merchandise, Gross Revenues
William H. Horstmann & Sons

Year	Coach Lace	Dress Trimmings	Foreign Imports
1860	$14,165.52	$ 10,001.84	$ 1,291.94
1861	$64,483.79	$ 87,113.93	$ 6,307.96
1862	$12,026.78	$ 7,441.48	$ 7,673.65
1863	$86,891.00	$114,385.00	$11,248.12
1864	$97,264.36	$167,033.11	$14,321.31

In 1858 the gross yearly cash totals of the three Philadelphia firms, including William J Horstmann's trimming store were:

William H. Horstmann & Sons	$86,594
Horstmann Brothers & Company	$33,147
William J. Horstmann & Co	$66,665
Total	$186,406

Factory balances for the year 1862, dated 1/1/1863, show the profitability of some of the functions the firms was engaged in. Reporting gross profits of $58,049.60, the figures were broken down as $11,610.88 in the sword account, $612.52 in the drum account, and $45,826.20 labeled under "sundries". These figures, the only ones which specifically show the profitability of the new factory functions started by the Horstmann brothers after 1848, show clearly that making swords was a profitable undertaking, and that drum manufacturing was not. The fact that only one workman is shown in this function in 1868 indicates that it was being phased out, and that most likely Horstmann was subcontracting his drum business by the end of the Civil War. Unfortunately, no mention is made of any separate account for Horstmann's insignia business. However, the fact that no new dies were created for this business after 1863 suggests that the firm had already begun to subcontract production of stamped metal ornaments, as they clearly did at a later date with the firm of Thomas Parry & Company of Philadelphia and which they had always done with the buttons bearing the firm's name.

The Horstmann Bros. & Co. balance sheet of January 1, 1864, shows a debit of $676 to the Cooper Firearms Manufacturing Company. While the Horstmann firm never dealt in firearms, this account statement, together with the presentation pair of Cooper's pistols noted as a show piece in Horstmann's exhibit at the Great Central Fair of June, 1864, in Philadelphia, makes it likely that Horstmann had some brief interest in the Cooper revolver. The Cooper, however, was a Civil War production which made no real impact upon the revolver business of larger firms such as Colt and Smith & Wesson. The venture was allowed to expire at the end of the Civil War. With this enterprise ended the one fleeting association of the Horstmanns with the firearms trade.

The firm's 1858-60 Cash Book is also deposited in the Historical Society of Pennsylvania. While this ledger does not cover the supercharged business conditions of the Civil War, it does document many of the standard practices engaged in by the Horstmann firms. Although the firm name recorded by the ledger is not given, it is almost certainly the original cash book reflecting purchases by Horstmann Brothers & Company. Its beginning date corresponds to the commencement of Horstmann Brothers & Company and the transactions that can be identified through this ledger are mostly for military merchandise, the division of the business allocated to Horstmann Brothers & Co.

The book is actually a "journal" or cash ledger in which all of the transactions for a certain period, including purchases, accounts receivable, and bills are given under the subtitles of "Debtor" and "Creditor," recorded using double entry bookkeeping. Here we get a unique view of the business relationship Horstmann had with specific firms, usually Philadelphia and New York business houses. In the credit accounts of February, 1858, Horstmann paid small bills for W. G. Mintzer, Henry Duhring and Co., T. N. Dale & Co., Klemm & Bro., and James Walton & Co. While all these are only a few dollars, if even that, the transactions reflect small orders from two military goods firms in Philadelphia, the largest silk trimming supplier in the Northeast, located in New York, a musical goods warehouse, and another military goods house.

The ledger also makes continued references to not inconsiderable purchases from William Pinchin. The account for July, 1858, shows a bill of $105.73. This is a matter of some importance, because Pinchin was George Armitage's successor in the military insignia manufacturing business. He occasionally tried his hand at running a military goods store. At this point, Horstmann was probably purchasing insignia to supplement its own line. Several additional Pinchin purchases are noted in the ledger. When Pinchin died in 1862, the Horstmann machine ledger records in its August balances, "Lot of machinery and tools belonging to estate of William Pinchin, obt. At sale at M. Thomas & Sons, $139.52." It is interesting to note that the amount paid for Pinchin's machinery and dies was not far in excess of the amount paid in a single order for his goods a little earlier. In all probability, Horstmann was simply buying a business which had supplied him over the years, with the intent of keeping his trade going. This ambition would seem to have been unfulfilled in the long run.

Also listed in the Cash Book are the following transactions:

J. M. Migeod, July 1858, $136.36
Cornelius & Baker, July 1858, $54.00
J. H. Lambert, July 1858, $1.66
D. Norcross, August 1858, ---------

Klemm, August 1858, $35.00
S. D. Walton & Co., August 1858, $292.05
Sullender & Pascal, October 1858, $7.50
W. G. Mintzer, November 1858, $84.00
William Pinchin, November 1858, $50.00
P. Wilson, August 1858, --------
George Hoff & Co., November 1858, $397.74
H. W. Conner & Co., February 28, 1858
Canfield Brothers, February 28, 1859, $30.32
Jacob Reed, April 27, 1859, $41.85
William H. Gray, June 19, 1859, $300.00
Freight on buttons from Lane, June 27, 1859
Thomas N. Dale & Co, June 30, 1859, $3.00
C. Diederichs, June 30, 1859, $3.25
Bartolett Blynn & Co, June 30, 1859, $533.50
Int. C. Oakford, June 30, 1859, $.20
J. Earle, July 21, 1859, $18.25
Kent Paine (int), October 24, 1859, $1.00
Kent Paine, January 1, 1860, $30.00
W. H. Gray for cross sabers, November 1859, $1.68 (12/2/1859 $2.68)
Mitchell & Tyler, November 16, 1859, (int $8.00)
Jacob Reed, November 30, 1859, $203.79
Sullender & Pascal, April 12, 1860, $6.46
Jacob Seeger, July 9, 1860 (remitted), $2.40
Pd. N. W. Bittes for Painting Flags, June 28, 1860, $9.00

These houses represent Horstmann's competition as well as his constituents. Migeod ran a small military goods business in Philadelphia. Cornelius & Baker were military goods contractors in the Civil War. Lambert was a hatter who ran a small military store in Philadelphia, while Daniel Norcross was a military goods dealer in San Francisco. Sullender & Pascal were also furnishing military headgear to volunteer militiamen, while Mintzer ran a fairly sizeable military store in Philadelphia. William H. Gray made trimmings such as epaulettes at this date, but later hilted and then sold swords during the Civil War. He ultimately made his fortune in casting gas fixtures. Philip Wilson was an arms dealer and subsidiary of P. H. Justice, of Philadelphia, a Civil War contractor and maker of heavy railroad equipment. George Hoff, another Philadelphia firm, was a clothier. Canfield Bros. ran a large military store in Baltimore with many customers in the South. Kent, Paine and Mitchell

& Tyler were military and hardware dealers in Richmond who would shortly supply the Confederacy with military items. Also recorded are bills for advertisements in the *Richmond Examiner* on April 23, 1859, and June 27, 1860. H. W. Conner of Charleston was a well known merchant who is noted in R. G. Dun and Company as financially backing Courtney & Tennent of that city, Confederate suppliers.[103] Charles Oakford was a well known Philadelphia military hatter while John Earle of Boston and Jacob Reed also of Philadelphia made large quantities of uniforms, some of which they may have supplied to Horstmann.

From these transactions we can begin to trace a web of business alliances between such cities as Philadelphia, New York, Boston, Charleston and Richmond. Horstmann was certainly buying from smaller manufacturers and businesses, and most likely he was selling to them also. However, because no Horstmann sales ledger has survived, that portion of the Cash Book which occasionally deals with interest on sales is very rudimentary, and cannot be used to document the same web of alliances on the sales end of the business as it does for the buying. However, given the business practices of the time it is not unlikely that such a wholesale web existed between Horstmann and smaller dealers who became the firms' customers.

One of the few specific transactions in the cash book shows Horstmann purchasing crossed sabers (cavalry insignia) from William H. Gray, additional documentation that Horstmann was subcontracting much of this production even though the firm had the ability to manufacture its own insignia in the late 1850s and was purchasing the dies to do so.

Horstmann's reliance on the Lane Manufacturing Company and on Scovill to make buttons sold through Horstmann's retail stores shows the way this aspect of the trade actually functioned. As described above, Horstmann never invested in the sophisticated machinery needed to make buttons, but purchased through one or the other of the large Waterbury concerns.

There are also ledger entries for painting and fringing flags. These entries raise the question of exactly which operations Horstmann was conducting in his "flag room". The large bills to Jacob Reed, a well known maker of uniforms in Philadelphia, and a smaller amount to John Earle may indicate that Horstmann was selling uniforms during the Civil War. A bill from Horstmann Bros. & Co. dating to the late 1860s to a captain C. P. Warner includes the notation "uniform made to order." This

[103] Bazelon & McGuinn: *A Directory of American Military Goods Dealers and Makers* (Manassas, REF Typesetting and Publishing, Inc., 1992). Entry H. W. Conner.

citation represents the earliest definite proof that Horstmann was selling garments. During the Civil War the firm was clearly in business selling everything except long arms and the actual jacket and trousers uniform set. While uniforms were later to be a mainstay of the company, the Horstmann firm seems to have avoided actual tailoring for most of its early history.

By the Civil War the firm was probably having an increasingly difficult time not selling uniforms when the mainstay of their military business was producing and selling uniform trimmings and retailing headgear. On the other hand, hats were clearly a long standing part of the Horstmann trade. A chapeau with a Horstmann label is in the collection of Andalusia, the Biddle family museum in Bensalem, PA. This hat belonged to Commodore James Biddle who died in 1848. While this is among the earlier Horstmann labeled hats that have been located, advertisements show the Horstmanns dealing in caps at least ten years earlier.

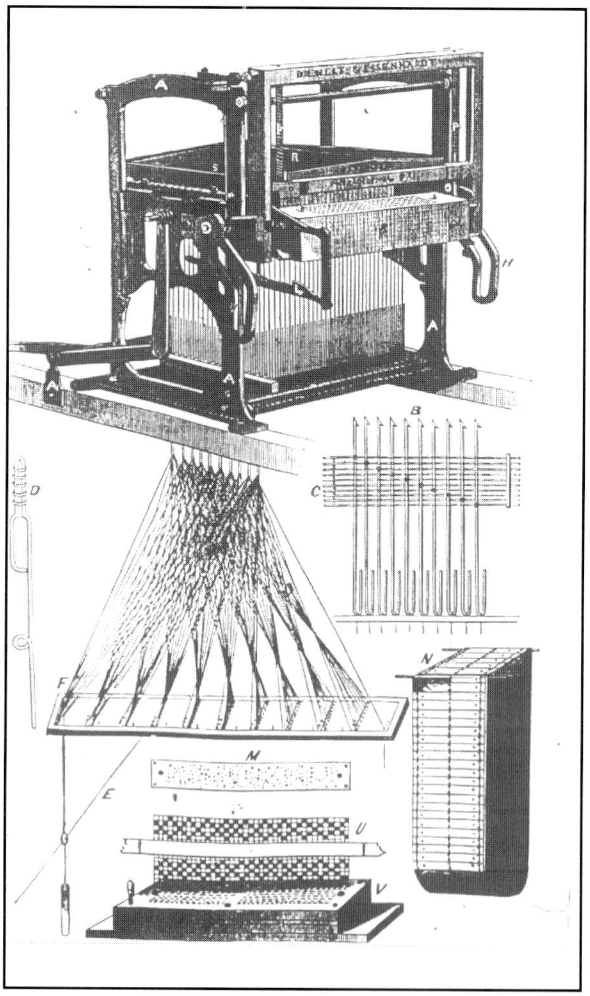

Drawings illustrating broadside explaining the use of the Jacquard loom. (Horstmann scrapbook c. 1860.) *See pp. 77-78 for the text of the broadside.* *It was W. H. Horstmann's introduction of this loom which gave his firm its initial leadership in technology.*

Facts of Interest connected with the Jacquard Machine.

Before the invention of the Jacquard Machine, the warp-threads in weaving figures were raised and lowered, in order to allow the shuttle to pass through, by means of treadles, which the weaver operated with his feet, and this method was in use for a long time; but as an operator could work but from 20 to 30 treadles, the figure or design was very limited. In order to weave a figure more elaborate it became necessary to construct some better arrangement. The first improved apparatus was made with cords, on which the design was placed and fastened on the side of the loom, and was operated by a boy. After this a Mr. Voncanson, from the city of Grenoble, in France, a well-known machinist, constructed a roller machine, with a tambor or drum, and this idea was probably taken from an organ or music box. On the tambor or drum he placed the design with pins, and these pins pushed the levers to raise the warp-threads. After that wooden cylinders were made, with from 20 to 32 flats or sides, on which the design was made by filling the holes in them with wooden pegs.

The last-named method, no doubt, gave the idea to Mr. Jacquard (a french weaver) to use paste board cards in place of so many flats or sides on a roller. He reduced the roller to eight sides, on which he made the design with pasteboard cards having holes in which he placed pins (to move the needles and hooks) according to the design. These cards he fastened together by cords, thereby forming an endless chain of as many cards as he required to complete the pattern or figure, and these cards he operated with a treadle in connection with the other machinery.

But as it is with all inventions, improvements follow improvements, so that the Jacquard Machine in use at present (although the same in principle as the original invention of Jacquard) has many decided improvements over the old style of machine, which can plainly be seen by the Improved Jacquard Machine (see cut) patented by Dienelt & Eisenhardt, of Philsaelphia, Pa., and for which they received a medal at the Centennial Exposition. No doubt the operation of these machines has been witnessed by thousands at the space of the Phoenix Manufacturing Company (B. B. Tilt & Son) in Machinery Hall, where they were used to weave those beautiful bookmarks and badges with the portraits of Washington, Lincoln and others woven in them.

Referring to the cut it will be seen that this improved machine shows a cast-iron frame A, in which are placed the hooks B, standing upright, and held in position at the bottom by the hookboard and on the top by the needles C. Needle C is held in position by the needleboard and spring box. On the outside end of the needles C is a spiral spring D, which presses the needles C, and through it the hook D forward through the needleboard and against the knives S, and with those hooks, needles and springs, in connection wiith the knives S, is it possible to raise the warp-threads E, as the warp-thread E passes through an eye or mallion which is connected to the harness F. The lathe or frame I, with cylinder K in it, is reciprocated through the roller G, running in slide or cam H; and at each forward motion of the lathe I, the cylinder K is turned forwards or backwards by the hooks L, and by this operation a card M, from the endless chain N, which passes over the cylinder K, will be placed and pressed in front of the needles C. The cylinder K is held in its right position by a hammer O, on which is a strong spring P, which presses the hammer O, down on the cast-iron cylinder head on the end of cylinder K. The perforated cards M, which are perforated on the machine V, after the design U being pressed against the projecting end of needles C, will push the hook B back where there is no hole; and where there is a hole in card M, the hook B will be lifted up by the knives S, held in the knife box R, and in this way the warp is lifted for the shuttle to pass through. The hooks B, which have been pushed back by the cards, will be pushed in their place again as soon as the cylinder K is passing forward, by the spiral spring D on the end of the needles, and so be ready for the next operation.

If, in operating the machine, there were no cards between the cylinder and the needles, then all the needles would enter the cylinder and all the warp-threads would be raised, and if, on the other hand, there were unperforated cards between the cylinder and the needles, then none of the needles could enter the cylinder, and consequently none of the warp-threads would be raised. It follows, then, that where warp-threads are required to be raised, it is necessary to perforate the cards as required by the pattern. Each card represents one passage of the shuttle with the material for the wool through the warp, and the number of cards necessary are regulated by the extent of the design. In the woven specimens of "Jacquard," "Jacquard in his Study," "Jacquard Loom in operation," "Washington and Lincoln," which are of the highest order of Jacquard work, many thousands of cards, and years of patient labor were required to prepare the cards for the loom; and that considerable artistic skill is required of this class of work must be evident to all.

Chapter 5. The Civil War Era to 1900

In this chapter we shall examine how the Horstmann firms functioned as one of the largest entities in the Philadelphia business world.

The Horstmann firm had contributed generously to the Mexican War effort. However, because of the shortness of that conflict, the actual war probably had only a slight effect on Horstmann's already good military business. However, by the Civil War the three Horstmann firms, excluding William J. Horstmann's restricted trimmings business, represented the largest military goods firm in the United States. William J. Horstmann seems to have taken on the role of representing the firm in its position of community leadership. As President of the German Society in the late 1860's he demonstrated his position as a leader of the German-American community in Philadelphia. Given William J. Horstmann's important business position at the time, this was only one of a variety of public honors he held.

In 1863 the U. S. Sanitary Commission, a private organization filling much of the role of the USO in later conflicts, embarked on a campaign of fundraising throughout various American cities. These fundraising celebrations were of a scope unprecedented in the American experience thus far. In fact they served as models for the great fairs which would be held in America during the next half century. The Sanitary fairs fulfilled the added function of soldiers' relief. They were organized and held in most larger American cities during 1863 and 1864. They all had in common an intense fundraising effort to benefit the Sanitary Commission. As fairs, they exhibited the current products of

American industry and displayed relics of American history.

Philadelphia's Great Central Fair was held in June, 1864, and was one of the largest, if not the largest, of the fairs. William J. Horstmann headed the military goods department of this fair, which was an important position during wartime. As such, he organized the support of notable military figures and manufacturers in the city. William G. Mintzer, H. W. Hensel, George O. Evans, G. W. Simons, Christian Sharps and his partner W. Hankins, J. P. Baldwin, and even out-of-towners like Joseph Cooper of Pittsburgh represented the military goods manufacturers on this committee. They included many notable military and social figures of the day.[104]

In addition to chairing the committee, Horstmann also put together one of the larger booths in the fair, complete with examples of Horstmann swords and a presentation pair of Cooper pistols. A bronze statue of a medieval knight on horseback imported from France at a cost of $150.00 was given as a prize at the Horstmann booth. This was by no means a unique presentation. The firm of Barton Jenks also had an impressive exhibit at the Great Central Fair. Jenks was chairman of the committee on arms and ammunition, and the manufacturer of Bridesburg muskets for the U. S. Government. As such, Jenks was one of the largest private rifled musket contractors of the Civil War.

The Ladies Committee on military goods, a part of Horstmann's committee, was chaired by none other than William J. Horstmann's sister-in-law, Mrs. Sigmund H. Horstmann, and Susan Trotter, another relative.

One interesting fundraising technique at the fairs was the donation of important presentation items such as swords, trophies, camp chests, and other honors which were voted upon by visitors who paid for the privilege of casting a ballot for their favorite military figure. The general with the highest total of votes received the presentation. The grand prize of the fair was a presentation sword donated by Evans & Hassall, valued at $2,000. This was awarded to General Meade after a hard contest with Pennsylvania Generals Hancock and Burney. The 177 votes cast for Grant were significantly below the 3,442 accumulated by Meade, and even the 1,506 obtained by Burney. At this point in the war Grant was not terribly popular, while Meade and Burney were local heroes. McClellan's fallen star was represented by his 297 votes, although he too was local to Philadelphia.

---

[104] Stille, Charles J. *Memorial of the Great Central Fair for the U.S. Sanitary Commission*. 1864. Also, the fair newspaper entitled appropriately *Our Daily Faire* published June-September, 1864.

Other prizes included presentation fire apparatus, a portrait of Meade, a camp chest, and a vase donated by F. Bailey. All of these prizes cost $20 a vote, and netted $10,457 for soldier relief, a good sum at the time. The fair included such things as a Hall of Flags and Trophies, and committees were formed to solicit exhibits from manufacturers of hollow ware, musical entertainment, tobacco, arms, and fancy goods, all of which represented different facets in the lives of well-to-do Philadelphians.

There is another side to the story of Horstmann and the Civil War. This lies in the character of business during the third quarter of the 19th century. New York, and to a lesser extent Philadelphia, functioned as the economic capitols of the South. When the South left the Union in 1861, New York Mayor Fernando Wood talked seriously of New York City seceding from the Union along with the South. The disaster of losing long-standing business friends created a brief depression in the Northeast which is well reflected in the R. G. Dun and Company ratings of this time. Individual merchants made their own decisions, especially if they had branches in the South. Ethelbert Halfmann of Philadelphia had opened a branch in Montgomery, Alabama, in the late 1850s and had gone so far as to sell his Philadelphia store, becoming, in the words of R. G. Dun and Company, "a strong secesh."[105] He apparently died in Havana in the middle of the war, waiting to smuggle in a load of goods, presumably military, such as the Confederate buttons that bear his mark.

Many large firms also had longstanding trade with the South, including military goods dealers such as Tomes Melvain and the Folsoms of New York, and B. F. Kittridge, whose business branches were located in many river towns along the Mississippi. While the Horstmanns had no branches in the South, their strong stand on the Mexican War does convey a certain indication of their feelings toward Southern expansion. In the absence of sales records, some concrete evidence of Horstmann's southern trade does exist. Many pre-war Southern state seal buttons have the back mark, "Horstmann Bros. & Allien," and others "Horstmann Bros. & Co." Each of these buttons represents a Southern order to Horstmann. The Horstmann balances carry a credit account with Hayden Gregg & Co., a prolific military goods firm from Charleston, and record other orders by the firms of Kent, Paine and Mitchell & Tyler located in Richmond. In addition, the dies created for the Horst-

[105] R. G. Dun and Company Collection. Pennsylvania Vol. 135, p. 59, cited in Bazelon & McGuinn: *A Directory of American Military Goods Dealers and Makers*, 1995 edition. p. 57.

mann Co. from 1859 through 1861, listed as purchases in Horstmann's machine book, show many insignia dies used to produce Louisiana belt buckles and crossbelt plates, a North Carolina buckle, and a Texas buckle, all in addition to the regular generic U. S. Army insignia. In February, 1861, only a few months before the war began, Horstmann made the punch and dies for a pelican cap plate, the seal of Louisiana, and in March the corresponding belt ornament. In April, 1861, the month Fort Sumter was attacked, Horstmann paid a bill for cutting and forging a die for the centers of North Carolina and Texas belt clasps. A number of these dies still exist, and they apparently have seen little use.

All of this simply reflects normal business relationships. After all, W. G. Mintzer ordered buttons, probably from Steele & Johnson of Waterbury, Connecticut, with the Georgia State seal, and dated on the back 1861, together with Mintzer's name and city. This was an obvious ploy to show that merchandise was ordered before the start of the war. Indeed, no other dated button is known to exist except for the state seals in this order, and similar buttons carrying the Steele & Johnson backmark.

In Horstmann's case, there is one other strange piece of evidence through the reminiscences of Levi S. White, a smuggler who specialized in transporting buttons and percussion caps from the North to Richmond. His story is told in four issues of the *Baltimore Sunday Sun*, from December 14, 1907 to January 5, 1908. White states that he had many Northern merchants who were willing, and even eager, to do business with him, knowing that the ultimate destination of the goods was to be the Confederacy. Apparently General Gorgas, the Confederate Chief of Ordnance and a Philadelphia native, told White that he had very few musket caps or potassium to put in them. White commenced his career by smuggling 500 pounds of chlorate of potassium which he had obtained from a Philadelphia druggist well known to Gorgas. He was also approached by a Baltimore die sinker and engraver to sell Virginia buttons which that individual had on hand. The die sinker referred to is probably Jacob Gminder, who most likely did not make the buttons, but ordered them in turn from firms in Waterbury, and then sold them through White to Mitchell & Tyler of Richmond. White occasionally employed an unusual method of shipping his goods south. This process consisted of White boarding a train with his bags of contraband in Washington and then jumping off before the train reached Harpers Ferry. He then made his way through Confederate lines, finally catching a Confederate train for Richmond. This is a dangerous feat if your cargo is volatile potassium chlorate. With all his jumping on and off trains, White could easily have ended his career as a small explosion.

White states, "In New York I could purchase almost as readily from a Union man as from a 'sympathizer.' No questions were asked us. One dealer on Maiden Lane [sold White] a large quantity of Virginia, South Carolina, and Georgia buttons although he had a son in the Union Army. 'We are here to do business. We have goods to sell, and our business is to sell to anyone who has money to pay for them. . . .' From another firm on the same street in New York I purchased more extensively. It was a branch house of a large manufacturing firm in Philadelphia. At my first dealing there I discerned that the resident manager was a strong Southern sympathizer. He bluntly told me that he knew for what purpose the goods were being purchased. 'But,' said he, 'I wish you success. I will sell you anything you may want, and if I haven't it I will get it for you.' He proved to be the most valuable ally I had through the war."[106]

While the first Maiden Lane firm was perhaps that of the Tomes Melvain company, the second and more important supplier could only have been Horstmann Bros. & Allien, located at 8 Maiden Lane from 1838 to 1868, the branch house of Horstmann's Philadelphia business. Whatever conclusion we draw from Mr. White's confessions, the Civil War fostered a generation to whom business was indeed business, and it would be unfair to hold the Horstmanns and Allien to standards which were not otherwise expected of them within their own trade.

The bulk of Horstmann's business now switched from a national peacetime trade to one of outfitting the Northern military effort during the Civil War. Reference was made above to the beginnings of Horstmann's contractual relationship with the Federal government. Horstmann's peacetime military customers included soldiers of the volunteer militia and the officers of the regular Army who were required to purchase their own uniforms and equipment. The enlisted personnel of the Federal services were issued arms, clothing and accouterments at government expense. Hence the government itself was the purchaser of items intended for issue. In addition, the government occasionally purchased items used by officers for which they were eventually charged. Issuing arms to officers remained an unusual circumstance.

The states also contracted for military equipments at the beginning of the Civil War. According to the militia act of 1792 the states were responsible to equip their own militias. This proved difficult to enforce and by the Militia Act of 1808 the Federal government took over con-

[106] *The Sun*. Baltimore, December 22, 1907, p. 12 col. 1.

tracting for arms, ammunition and essential equipment such as the cartridge box. The volunteer militia, it will be remembered, purchased its own uniforms and equipments though it supposedly received arms from Federal arsenals.

The Federal procurement system was in theory based upon the production of arms for the regular Army at Federal arsenals, while contracts for arms would be used to supplement the arsenal supplies. This proved a good deal more complicated in practice, with needy units often being issued the arms at hand regardless of their status. The system was extended to the manufacture and issue of clothing and ammunition at Federal Arsenals such as the "Clothing Laboratory" at Philadelphia's Schuylkill Arsenal and the cartridges produced a few miles to the north at Frankford Arsenal.

The obligatory militia maintained by each state had become ineffective, and in fact was neither mustered nor uniformed by the states. Its officers were usually the only uniformed soldiers of this non existent army and they purchased their own uniforms. Still, the states maintained a large number of Federal arms which were to be issued as needed. Thus, by the mid-19th century the obligatory militia had become a paper force which never met in practice but which could be called out in an emergency.

Rather than activate the unwieldy obligatory militia, at the beginning of the Civil War the states were given quotas of volunteers to raise for the Federal war effort. Many of these troops were raised from the existing volunteer militia, whose soldiers enlisted as whole units or formed the cadre of experienced soldiers in a variety of different regiments. Facing needs to issue equipment on an unprecedented scale, the states attempted to contract to equip their troops through the year 1861. The system had become so tangled by the end of that year that the Federal government took over the entire contracting process in January 1862, including the obligation to fully clothe and accouter as well as arm the troops under their direct jurisdiction. The Federal government had managed the same way in 1812 and in the Mexican War when state troops fought under Federal jurisdiction.

The Horstmanns had continued their contracting for the Army through the 1840s and 1850s. These contracts are duly reported in the U. S. House and Senate documents issued each year to account for the expenditure of public funds. Their 1853 orders included such items as 200 sashes, 100 drumsticks with their "carriages", pompons, buttons, and the traditional 1/3″ worsted cord in fairly large quantities. The contract of July 8, 1853 included 2800 yds. of orange, 52,000 of scarlet, and 102,000 yds. of dark blue corresponding to the colors worn by the dragoons, artillery, and infantry respectively. In 1856 they supplied such items as 400 yds orange silk ½″ wide, 800 green, 1800 scarlet, 1000 blue, 50 drum slings at 37½ cents, 1400 gross brass coat buttons at 65 cents per gross, 900 gross vest buttons at 45 cents per gross, cords, and tassels. By 1858 they supplied 1050 gross coat buttons and 1250 gross of vest buttons at the same prices as in 1856, 25 shears ferules for guidons, 135 drum slings and 200 drum heads. While not a complete list, these products are typical of Horstmann's 1850s contracting business with the Army. Because the regular military forces were small at this time, these contracts can only have represented a fraction of Horstmann's total military business, the majority of which was probably done with state militia organizations.

All this changed with the Civil War. Typical orders for Horstmann Brothers & Co, the firm which nominally carried most of the Horstmann military business from 1858 to 1876 are:

8/19/61	12,000 yds blue worsted
2/12/62	50 national colors $55.00 each
2/21/62	500 drum heads and snares
9/1/62	1250 common tents $2.50 ea
9/9/62	10,000 flannel drawers
10/8/62	20,000 yds 3/8 inch yellow worsted lace 1.5 cents per yd
10/8/62	200 infantry drums $5.25 ea
11/3/63	100,000 yds 3/8 inch scarlet lace and 200,000 yds yellow
6/20/63	500 cavalry guidons
7/7/63	2000 pr. drumsticks
7/3/63	2000 bugle cords and tassels
11/6/63	100 infantry bugle cords and tassels
11/20/63	1000 drum snares, 10,000 brass corporal, privates scales
9/21/63	1000 drums with sticks, slings and carriages
8/7/63	10,000 yds ½ inch silk lace yellow, blue; 500 hospital steward chevrons 40,000 yds scarlet lace and 10,000 brass castles
4/22/63	60,000 brass numbers
8/3/63	1 inch webbing 420 yds

10/19/63	worsted sashes
9/28/63	1000 drum sticks and carriages
10/12/63	4000 bugles
11/9/63	200,000 yds ½ inch worsted lace
11/9/63	1000 trumpets, plain
4/20/64	1250 infantry chevrons cavalry, 480 c. Sgt, 25 sgt arty silk, 2500 first sgt, 25 qm sgt silk, 300 inf, 10 qm sgt, silk
6/26/64	500 infantry drums
8/26/64	300 bugles
9/14/64	50 national colors
9/11/64	600 worsted sashes
9/26/64	194 corps and 650 brigade flags
1/14/65	10,000 yds 3/4 inch yellow worsted lace
1/28/64	374 brass eagles, 1800 red cords and tassels
4/12/64	500 ambulance guides

In addition the firm of Horstmann Brothers & Allien were also contracting for much the same merchandise. Typically:

8/12/64	50 garrison flags
1/28/64	500,000 yds ½ inch blue worsted lace
9/28/64	700 cocoa wood fifes
10/27/64	3000 pair drumsticks
2/3/65	2000 yds scarlet lace worsted

For instance, Horstmann is known to have supplied, along with Evans and Hassall, virtually all of the colors used by Pennsylvania's 202 regiments, each of which would have had at least one issue of two flags.[107] The above list does show the diversity of products which Horstmann had undertaken to supply. Like many other contractors of that era, Horstmann produced some items itself and subcontracted the remainder. Even where Horstmann made items such as flags, the employee register cited above shows the firm sending out work of applying fringe to and painting their flags. The firm was clearly purchasing its sword blades as finished components, and may have contracted some of the sword hilts and scabbards during the war when demand certainly outstripped the firm's capacity. No new dies are listed in Horstmann's ma-

[107] Richard A Sauers: *Advance the Colors* (Harrisburg: Capitol Preservation Committee, 1987). 2 volumes.

chine purchase register to carry his heightened wartime demands. Buttons were certainly not produced at the Horstmann factory, but some limited quantity of the insignia it sold probably was made by Horstmann during the war years, as the firm had already made up the dies to produce most of the standard devices called for in Federal regulations of 1851 and 1852.

Most surprising is a list compiled by Ron Hickox from the National Archives Ordnance files RG 156. He shows that over 16,000 cavalry sabers were ordered from Horstmann, approximately 4000 NCO swords, 5500 musicians' swords, 400 artillery sabers, and 87 heavy artillery sabers were likewise supplied. Given that Horstmann was also supplying many of the officers' swords used by the Union during the war as well as the high grade presentation swords, the firm's sword trade was doing well indeed. Except for the part ownership of the Cooper revolver described earlier in this chapter, Horstmann never seems to have contracts for or supplied firearms. He could, however, cooperate with firearms manufacturers. Some rare brass hilted bayonets are marked Horstmann and feature a peculiar blade with the hilt in the form of a brass eagle. This bayonet was made for a rifle musket supplied by John Wurflein of Philadelphia, probably to a single unit.

As a firm, the Horstmann companies had no political opinion. The founder "... was a staunch upholder of civil liberty and freedom of opinion. Although deeply interested in national and state problems, he was no politician. He was eminently practical in everything, took little interest in polemics and had scant patience with demagogues and theorists ... and he frequently expressed himself in terms of warm approval of the splendid advantages open to all classes alike in this favored land of America."[108] Given their support for the Mexican War, the Horstmanns may have favored the party of Jackson, the Democrats. However, on one issue the firm later evidenced strong opinions, namely the tariff question.

The tariff issue in the 19th century pitted the growing industrialism of the North against the agrarian South. The manufacturers of the North naturally wished to protect their growing industries, while the farmers of the South preferred cheap imports. As both manufacturers and importers, the Horstmanns must have had some mixed opinions as to the tariff rates on their imported goods. These rates went up or down but the normal 35% tariff on military articles seems to have favored the business.

Among Horstmann's normal imports were sword blades, an item

[108] *One Hundred Years*, pp. 26-7.

exported worldwide by the blade makers of Solingen, Germany. While the United States had several well known producers of blades, many of swords were in fact hilted in the United States using Solingen blades. One document in the author's collection shows J. E. Bleckmann, a manufacturer and exporter of hardware, tools, cutlery, and swords in Ronsdorf in Solingen, shipping "25 Blades No885 $1.75" (totaling $43.75) through its New York agent G.A. & E Scalded to none other than the Ames Manufacturing Company in Chicopee on February 17, 1862. Ames, the largest producer of swords, including blades, in the United States, evidently still had to supplement its production with orders from Solingen. Horstmann's import of German blades was thus a standard practice of many businesses and the tariff rates charged for the import of such articles were an important consideration for the entire trade.

The Horstmann scrap book contains an article dated July 13, 1870, from a newspaper entitled *The World.* This article charged that while finished swords were taxed at a rate of 35%, blades alone were subject to a duty of 45%–a situation which was most injurious to Horstmann who had to compete with completed foreign swords on one hand. On the other, he imported Solingen blades to be hilted in his factory and was thus penalized doubly by the higher rates on sword parts. The article satirizes U. S. Senator William Kelley of Pennsylvania as "Horstmann's son-in-law", a situation not literally true, but as a staunch protectionist Kelley was spiritually within the Horstmann family. ". . . For four years did the Jethro of Philadelphia [Horstmann] try to persuade his son-in-law [Kelley] . . . to reverse the order of things, viz. To make sword-blades 35 per cent and complete swords 45 per cent which little arrangement would allow the father-in-law to import blades from Birmingham, and make swords in Philadelphia; while he now has to buy blades in Pittsburg (sic) and make them up in Philadelphia. . . ." While there is no Horstmann comment found on the article, it is just as likely that the accusation that the firm would import and use anything British was the more hurtful of the allegations in the satire.

In fact there is indirect corroboration of this article in Lewis Heyl's compilation of tariff laws.[109] This digest shows the tariff of March 2, 1861, as having fixed a relatively low 30% duty on most clothing and metal goods. The act of July 14, 1862, added 5% duty to most of these items including "side arms of every description".[110] The specific problem re-

---

[109] Lewis Heyl: *United States Duties in Imports Published in 1874* (Washington: W. H. & O. H. Morrison, 1874.

[110] *Ibid.* Paragraphs 137, 297.

ferred to in the *World* article came with the act of June 30, 1864, which placed "... on all manufactures of steel, or of which steel shall be a component part, not otherwise provided for, forty-five per centum ad valorem. ..." This problem did leave Horstmann paying a duty of 45% on blades as components but only 35% if he imported complete swords. This problem was redressed in the Act of July 14, 1870, which set sword blades at 35% and swords at 45%, making things right between the government and Horstmann once again. In 1872 the duties on all metals were reduced 10% which may have assisted the situation still further, as imports of entire swords at this time were negligible and probably did not aid Horstmann's competition. Heyl also shows duties ranging from 30% on musical instruments and 35% on muskets, rifles, and other firearms, to a whopping 60% on silk trimmings, laces, and woven gallons. Imports of gold and silver embroideries, effecting the Horstmanns' trimming business, were fixed at 35% during the period.[111]

For our purposes the interest of the *World* article is its documentation of the sword business and the cryptic note that Horstmann's swords were made in Pittsburgh. Indeed there is some artifactual evidence for this, in that some of the blades on the staff swords of the last quarter of the 19th century show an Old English letter "H" on the ricasso of the blade rather than the usual Solingen marks of a king's head used by Weyersberg or the Clauberg companies. We cannot pursue the matter too far. One contemporary source continues to note that the "... blades are almost all imported from the ancient sword-blade emporium of Solingen. ..."[112]

In the absence of actual Horstmann production records and considering the existence of Horstmann swords which have the Solingen "knight's head" on the blade as well as the Old English "H," it is at least possible that Horstmann was developing a domestic source for his blades to replace his traditional reliance on Solingen-made products. The 1875 *Manufacturers of Pennsylvania* goes on to note "... a stock of some thousands of blades, of many different patterns and sizes, ready to be set and finished. Any style of sword can be had from this warehouse, from the plainest kind up to a presentation sword worth a thousand dollars. ..."[113] The Horstmann letterhead of the period featured a standing knight reminiscent of the

---

[111] *Ibid.* Paragraphs 363, 599, 719, 1111, 1264.

[112] *Manufacturers and Manufactories of Pennsylvania of the Nineteenth Century* (Philadelphia: Galaxy Publishing Company, 1875), p. 407.

[113] *Ibid.*, pp. 407-8.

Solingen blade marks with which the firm was so familiar.

Through the 1870s the Horstmann firm retained something of its German character. We have seen that the narrow weaving trade which was the core of Horstmanns' manufacturing expertise was practiced in America largely by immigrants from the Continent rather than from England. Horstmann's "boys" were given a German/American education at some trouble to their father. Upon graduation William Horstmann was at some pains to take his elder son to Europe to visit the places of his heritage. This later bore fruit with the continuing interest of William J. Horstmann in the German Society in Philadelphia, of which he was President in the late 1860s. During his presidency, William J. Horstmann wrote about important German scientists like Alexander Von Humboldt and German culture for the Society's newsletter. He did not write about Germany's military history. In fact Germany at this time had yet to acquire its reputation for military prowess, that honor going largely to France and Britain before 1870. The Horstmann scrapbook includes an article on relief for the French victims of the Franco-Prussian War in 1870, with William J. Horstmann's name prominent in the list of contributors to the French relief effort. While there may have been a cultural bias within the Horstmann firm, German militarism, as later generations understood the term, was not a part of the Horstmanns' heritage or their outlook on German culture.

Imports of machinery and component parts to be assembled in the Horstmann factory likewise came from the Continent and the register of employees discussed above has a predominance of Germanic names, particularly among those who stayed with the firm for some time. The early business partners in New York, Gustavus Drucker and Henry V. Allien, seem to have carried on a business connection with families having Continental origins in Europe.

However, this German-ness can be pushed too far. When it came time to marry in the late 1840s, Sigmund Horstmann married Elizabeth Cowgill West, while William J. Horstmann's bride was named Emma Walker.[114] Sigmund's daughter, Elizabeth Trotter Horstmann, did even better, marrying Walter Lippincott, and thus positioning the Horstmanns with Philadelphia's elite. None of these alliances strengthened the family's German ties.

---

[114] *Encyclopedia of Pennsylvania Biography* Vol XVII, Ernest Spofford ed. (New York: Lewis Historical Publishing Co, 1928), pp. 189-191. Entries for Walter Horstmann and Walter Lippincott. Also: Leonard, John W.: *Who's Who in Pennsylvania* (New York: L. R.Hamersley & Co, 1908), pp. 376-377.

Of the names rising to prominence within the company, certainly the first junior partners taken in by the firm–George O. Evans and William Hassall–are not German names. Other members of the firm prominent in the 1870s were Albert Weihenmayer, described in his obituary in the Horstmann scrap book as a German, who came to Horstmann about 1853. Samuel Clarkson would later emerge as an important figure in the firm. He is described as William J. Horstmann's brother-in-law, joining the firm in 1867.[115] J. Howard Mecke was a member of the Union League and an Episcopalian, not strong German credentials at the time. On the other hand, George and William Schultz Jr. most likely shared the Horstmanns' ethnic heritage. Other long time employees of the time–Albert Cullom (a drum maker), John W. Firth, Edward L. Calely, and John Frasier–seem not to have shared the origins of the founders.

As the 1870s dawned, William J. and Sigmund H. Horstmann had every reason to look forward to a bright future. Well established, the business was hugely successful and could look forward to more years of expansion under their wise leadership. William J Horstmann, always the front man for the brothers, was given a prestigious appointment to the United States Centennial Committee created to bring the Centennial celebration of 1876 to Philadelphia. All was going very well. It was not to be.

Sigmund Horstmann's health was somewhat delicate, ". . . having suffered for years from a painful disease . . ."[116] More importantly, he had a sickly daughter, Sallie, who accompanied Sigmund and Elizabeth Horstmann to Rome in March of 1870. A description of the tragic events of the trip, found among her papers, was written by Elizabeth Horstmann in memorial to her husband and daughter. Sallie contracted typhoid and died on March 23, 1870. A week later Sigmund also had the disease and died in Rome on March 29 at the age of 49.

William J. Horstmann inherited Sigmund's half of the factory together with his ". . . half part of all the steam engines, shafting looms, machinery pipes and gas fixtures. . . ."[117] It apparently meant little. The brothers had always worked so closely together that ". . . each looked upon the other as another self. . . . William always carried this grief in his heart, and it soon became apparent that his health was failing. He

---

[115] Philadelphia City Archives. Will Book 74, Entry 383, p. 556, William J. Horstmann, May 28, 1872.

[116] *One Hundred Years,* p. 27.

[117] Philadelphia City Archives. Will Book 69, Entry 650, p. 228.

visited California in 1872, and seemed to suffer more from mental and physical dejection than from any other cause. He was unable to rally after his long railroad journey . . ." and died in California May 10, 1872, age 52.[118]

We can view the remaining family through passenger indexes showing several trips to Europe in the mid 1870s. The most extensive of these, aboard the steamer *Ohio* and dated September 28, 1874, shows Mrs. William J. Horstmann, age 50, Walter Horstmann, age 18, Lulu Horstmann, age 5, and Florence Horstmann, age 20, along with Mrs F. Oden Horstmann, age 26, Ethel Horstmann, age 2, and F. O. Horstmann, age 28. The latter listed his occupation as "regalia Mnfr".[119]

With William J. Horstmann's death the Horstmann companies passed to the third generation. Still a family partnership but no longer a simple business of father and sons, it was a huge enterprise employing hundreds. However, the dynamic leadership which had built the business had gone. The firm would remain very profitable, but the task was now one of running an existing business, not growth. Horstmanns had reached middle age.

William J. Horstmann's will noted agreements with Albert Weihenmayer, William Schultz, Jr., George H. Schultz, and Samuel Clarkson. It also provided for a happy arrangement with William J. Horstmann's sons, F. Oden Horstmann and Walter Horstmann then a minor, and with Sigmund's son William H. Horstmann (II).[120] Care was taken to preserve the working dynamic of the firm by including all of the principal managers in the firm's new management structure: ". . . the subject is a delicate one and I confide it to their judgement to continue the same unaltered. . . . I hope and feel they (Weihenmayer, the Schultzs, and Clarkson) will be dealt with liberally and that they together with my sons and my nephew William H. Horstmann become owners of the business." The children all received substantial (for the time) cash bequests, the boys $50,000 and his daughters Clara and Florence $25,000 each with additional sums to come to Florence after William's widow's death.

At the time, the value of the two firms is given in an 1869 R. G. Dun and Company entry as $1,220,000 which included dwelling houses valued at $90,000, stores and factory $870,000, $150,000 inventory, and a

---

[118] *One Hundred Years,* p. 27.

[119] National Archives. Philadelphia Branch. *Passenger Index, Philadelphia 1800-1906.* Microfilm reel M 360.

[120] *Op. cit.,* p. 27.

New York store at $100,000.[121] These figures are, of course, informal estimates but they do give an idea of the size and value of the business.

This arrangement worked for the next four years. On December 31, 1876, the New York branch of the firm, Horstmann Brothers and Allien, was dissolved and Henry V. Allien withdrew his interest in Horstmann Brothers & Co in Philadelphia as well. The New York branch was replaced by H. V. Allien & Company which, descending in the hands of Allien's children, lasted through 1948. This reorganization was immediately followed by the withdrawal of Sigmund Horstmann's son, William H. Horstmann (II) on May 18, 1877. William H. Horstmann (II) seems to have been associated closely with the New York operation and his withdrawal, probably for health reasons, may have been a factor in the Horstmann Brothers & Allien break up. In any case William H. Horstmann (II)'s health gave out two years later and he died January 1, 1879. With these departures, only the sons of William J. Horstmann retained control of the Philadelphia firms, William H. Horstmann & Sons and Horstmann Brothers & Company.

The new Allien firm consisted of Henry V. Allien, the Horstmanns' old family friend and partner; his sons, Laurent Allien and Louis S. Allien; and Nelson Thompson, a figure left over from the Horstmann days. In spite of their separation, it is very probable that the two firms continued to work closely together. H. V. Allien's personal copy of the 1877 Horstmann catalog exists indicating that Allien continued to sell Horstmann merchandise and utilize the same sources of supply as he had previously.[122]

F. Oden Horstmann (8/4/1846-10/18/1894) succeeded to his father's place as front man for the Philadelphia firm, though not to his role on the U. S. Centennial Commission. Oden Horstmann's role was more modest. As an important business leader he became president of the Silk Association of America, a trade organization dedicated to furthering the interests of the U. S. silk industry. In this capacity F. Oden Horstmann solicited the silk trade exhibits for the U. S. Centennial of 1876, continuing at least a part of the family association with that fair. Oden Horstmann continued to manage the business competently and from all available evidence, profitably.

The Horstmanns' association with the silk trade went back to the

---

[121] R. G. Dun and Company Collection. Philadelphia. Vol. 4, p. 408. Entry January 23, 1869.

[122] Author's collection.

earliest days of the industry in the United States. A large portion of the Horstmann scrapbook of the time is given over to articles about silk which sometimes document the role of the Horstmann firm in the burgeoning trade. An article from *Haddock's Sunshine* of February 10, 1877, taken from the scrapbook, describes Horstmanns' leadership in the trade, manufacturing a large range of narrow silk products and its competing cooperatively with firms in Patterson, N. J., and New England through the trade association presided over by F. Oden Horstmann. A similar article in *The Record* of March 26, 1885, shows Horstmann and other firms rapidly and quietly taking over the trade as early as the 1850s when the article notes that "silk trimmings and military and society goods are no longer imported in the finished state." The Horstmann firm had won a Franklin Institute award in 1852 for silk goods and an American Institute award for silk in 1869. Renditions of these medals were proudly displayed on the masthead of Horstmann's business letters of the early 20th century along with medals from the Centennial Exposition of 1876 and the National Export Exposition of 1899. Such awards were greatly valued by the 19th century manufacturer and were keenly sought.

The silk industry was then approaching its heyday, with large firms of at least the capacity of the Horstmann company located in Patterson, New Jersey; Scranton, Pennsylvania; and New England. One direct military competitor was T. N. Dale of New York, whose name appears on military buttons of the period but whose concern was really a huge cloth goods manufactory. In Philadelphia alone, there were 44 establishments making silk in 1860 with a product given in the *Philadelphia Record* of 3/28/1885 as $1,933,725.[123] This source gives the Horstmann firm as the oldest and largest of the Philadelphia firms making ". . . hosiery . . . Columbia Zephyrs, Germantown yarns with Saxony . . . worsteds." The Horstmann scrapbook's *Evening Bulletin* article of August 30, 1873, states that Horstmann employed 350 hands, 100 employees in sales, and 20 salesmen in Philadelphia, New York, and 38 Rue Meslay in Paris.

Post-Civil War business was good to the Horstmanns. The 1870s were a time of economic ups and downs, but this apparently did not hurt Horstmann's trade. According to a scrapbook article taken from the *Philadelphia Public Ledger* of August 28, 1877, the firm continued to employ 400 hands right through the depression of 1877. The same

[123] Horstmann scrapbook.

source goes on to say that the silk trade in Philadelphia was generally growing or at least holding its own. Merrill, Colladay & Co on N 4th employed 1000 people; J. C. Graham at 607 Cherry 527; H.W. Lipper on N 5th 600 hands, a 200 person increase in a few weeks. William H. Horstmann & Sons stood at 400 less than had previously been employed but "... this number is greatly reduced by improved machinery which they use. ..." The firm was functioning at full capacity with no reduction in wages. This was described as better than at any time since the war. The *Textile Register* of Jan. 1881 (Philadelphia, Horstmann scrapbook) stated that Horstmann was more profitable in the previous year than ever before. Salesmen were out all the time and the firm was making preparations to increase their production and markets. The Public Ledger of August 28, 1877 describes business as "... better than at any time since the war ..."[124]

The *Evening Telegraph* article of October 9, 1888 includes testimony by Samuel Clarkson in a case involving John Wanamaker on the tariff duty required for silk ribbons. In his testimony, Clarkson describes the Horstmann association with silk. He stated that Horstmann "... is engaged in the manufacture of narrow silk fabrics, and also imports military goods ..." and that "... the house has imported since 1820. ..."[125] The *Evening Bulletin* of August 30, 1873 notes the complexity of the Horstmanns' manufacturing operations and notes improvements to 5th & Cherry in the early 1870s. These improvements show emphasis at that time on the civilian end of the business and mostly on silk production, with a new store being added to the plant. The complex was now 50 x 144 feet on Cherry with buildings 22 x 144 feet turning the corner on Fifth. The firm was praised for its open skylight and attention to ventilation. Even the basement was lit by glass mounted in the floor above. The complex used hot water heat. The basement also held embroidery, silk, and materials in which Horstmann is cited as doing the largest business in the United States. Principal goods being sold were Berlin zephyrs and domestic yarns manufactured in the establishment. These were located on upper floors of the Horstmann factory along with upholstery trimmings and hosiery.[126]

The era of the Horstmann Brothers from 1845-1870 was one of experiment and reorganization. The company of the third generation of

[124] Horstmann scrapbook.

[125] *Ibid.*

[126] *Ibid.*

Horstmanns represented maturity and solid growth, the refinement of an already established trade.

As we reach the mid-1870s, the original sources which shed light on the inner workings of the company end. We no longer have cash, labor, and machinery records to compare to newspaper and other published accounts, so we must rely solely on secondary accounts once again. However, one source, the Horstmann scrapbook, continues well into the 1890s and is still highly useful. By 1877, the firm also began to issue its own catalogs of goods. These catalogs represent a major source of knowledge as to what was being sold, but do not reveal a great deal of the inner workings of the business.

Comparing the Horstmann trade catalogs to those of other large companies in the trade shows that even as late as the 1870s the company did not yet rely on uniform clothing, but on the accouterments, headgear, and ornament associated with the military and fraternal trade, and with trimmings rather than finished cloth in the civilian end of the business. The 1877 catalog shows "Equipments for Uniforming the Army and Navy and the National Guard, Bands, Clubs and Firemen". These equipments included swords, belts, insignia, epaulettes, chapeaus, spurs, buttons, gauntlets, bindings, cords, plumes, drums, laces, sashes, facing cloths, embroideries, fifes, trumpets, bugles, chevrons, hat cords, etc. The styles shown in the catalog are a strange mixture of the current, and pieces which should have gone out of fashion in the 1850s. Prussian helmets of the 1872 Army regulation are shown together with Civil war kepis and Hardee hats of an earlier generation. The swords depicted reflect current regulations together with regulation of 1851 swords, deleted by the Army in 1872, and the still older Widmann types for which Horstmann had acquired the tools in 1848. Buttons also depict old patterns mixed with the new.

This mixture of styles reflected traditional business technique. Before the Civil War, city militia companies often sold their old uniforms to more rural units. However, given the plethora of bands, theatrical, and fraternal uniforms which abounded after the Civil War and which were less fussy about current Army regulation dress than the militia, a retailer like Horstmann must have felt that even old stock had its customers in these new organizations, though exactly who in 1877 was buying specifically marked pieces such as Army Topographical Engineer buttons which had gone out of regulation in 1862 does strain the imagination. Horstmann's 1888 catalog dropped many of the older styles and concentrated more on current goods, perhaps signaling an increased popularity of military service among Horstmanns' customers. This newer catalog puts less emphasis on miscellaneous headgear types and insignia and more on uniform parts, including the new Army regulation overcoat. However, aside from this one garment, the catalog still does not advertise uniform tunics or trousers, stopping at such non-tailored items as leggins, gauntlets, and headgear.

Looking to Horstmann's competitors for a moment, a Ridabock catalog of the mid-1880s shows almost the exact same distribution of goods, including the combination of old and new sword types. The Ridabock firm had been established by John A. Baker in New York in 1847 and had become Baker & McKenney during the Civil War. Its 1880s catalog also depicts a combination of old and new goods similar to that published by the Horstmanns. Mitchell C. Lilley & Company's 1882 catalog shows complete uniforms, their specialty, as do the catalogs of Pettibone & Company of Cincinnati. Pettibone and Lilley were quite versatile and sold such items as lodge furniture and other paraphernalia. Lilley had been an Odd Fellows book binder in his early days and got into the business from the fraternal ephemera side of the trade. Unlike Horstmann, Lilley did sell uniforms by the 1880s. George Evans & Co. of Philadelphia, started by a different Evans than Horstmann's junior partner of the 1850s, sold a variety of equipments, and also uniforms for military and band organizations. Evans also sold uniform coats and trousers by the 1880s. John Boylan & Company of Newark and later New York, together with its successors, W.C. Boylan and the Boylan Manufacturing Company, specialized in military clothing. Their 1884 catalog is addressed primarily to the different units of the New York National Guard but also includes Army, Navy, and police uniforms.

W. A. Raymold, a derivative of Raymold & Whitlock founded in 1881, and also of New York, sold military uniforms along with accouterments and swords. The firm of Frank Henderson of Kalamazoo, Michigan, begun in 1869, established perhaps the largest of the uniform manufacturing houses. Its successor, Henderson-Ames, sold uniforms, swords, insignia, horse equipment, and camp gear, as well as a variety of fraternal uniforms and accessories. This firm later merged with the Ames Sword Company of Boston, founded in 1881, and later with M.C. Lilley, creating the Lilley-Ames Company, which would absorb the military portion of Horstmann's business in 1923.

A catalog issued by the Ames Sword Company about 1882 is marked on the outside cover "Horstmann Bros & Co. Philadelphia, Pa," Horstmann's pre-1893 military business. The Ames Sword Company was founded in May, 1881 and the catalog dates shortly thereafter. Inside the firm takes issue with Horstmann's sword business and it is therefore surprising that Horstmann would have sold directly from this catalog. Evidently they were not sensitive to: "The Ames Sword Company are to-day the only real manufacturers of swords in this country. One or two companies import blades which they have mounted and sell as their own manufacture, but they cannot be considered, in the proper sense of the word, manufacturers."[127] While unflattering, the description of Horstmann's sword business, and of Widmann's before it, is nevertheless true. The catalog testifies to the diversity of fraternal swords being made and sold as well as the continued military business done by this newly formed derivative of the original Ames Manufacturing Company.

In addition to the military goods and uniform dealers, the immediate post war trade saw a new entity arise, the military surplus dealer. This business was created by the huge stocks of war material assembled by the Government during the Civil War and sold off after the end of that conflict. Of these dealers, the firm of Francis Bannerman is best known, but his was far from the only firm which bought and sold used goods. Horstmann's playing field in military, fraternal, and civilian items had become most crowded with firms as large or larger than Horstmann. They in turn competed with a large variety of smaller retailers, tailors, and dealers of all sorts.

Horstmann's customers were also changing. Before the Civil War the military unit was based upon the company, a unit of up to 95 men. Many of the companies formed in cities across the United States could afford their own peculiar styles of uniforms and accouterments. During the 1850s some of these companies in larger cities had become battalions of two or more companies. This was the precursor of things to come. After the Civil War, the standard military unit was the regiment of perhaps 1000 men, made up of perhaps ten of the older companies. This change in organization had its effect on styles. Instead of having a plethora of uniform fashions, each regiment had its distinctive dress worn by many men instead of a few. Even these styles tended to differ in insignia types and detail of trim rather than substantive differences in uniform. It was thus more practical for each of these units to have its own insignia instead of relying upon stock patterns, as had been the practice with less wealthy companies before the war.

---

[127] *Illustrated Catalogue and Price List of the Ames Sword Company, Chicopee, Mass. U. S. A.* No other information. The catalogue has over 386 pages of swords, belts, and buckles. Joe Marino collection.

Fraternal organizations had also grown dramatically. So many men were now involved in military and fraternal organizations that supplying them had become big business indeed. In order to get ahead, society encouraged its members to join something, and as the types of organizations multiplied so did the firms which supplied their needs. Significantly, many of the manufacturers began as members of the organizations they would outfit. Some of the smaller firms actually had space in fraternal lodge halls and grew out of some member responding to the opportunity to make a living from his membership. Even larger manufacturers found it prudent to become members of multiple organizations. However, Horstmanns, as a large and firmly established business, apparently did not respond to this dynamic of joining their customers. This approach was left to latecomers.

The Horstmann firm had always faced a competitive market. In the early days other Germans such as Henry Korn were making the same items as Horstmann in Philadelphia and certainly there were more than enough firms in the import trade. What had given Horstmanns the edge, at least through its first two generations of management, was a willingness to be bold and dynamic. This meant taking chances by importing Jacquard looms before the rest of the trade or adding new types of manufacture and sales to become the first military "department store" on a grand scale utilizing their state-of-the-art factory complex located at Fifth and Cherry.

Something had changed with the advent of the third generation. The Horstmanns still enjoyed a large share of a large market and probably took in more revenue than ever before but they were no longer the undisputed trade leaders; they were one of the pack. Accordingly the fortunes of the Horstmann firm would now rise and fall with general market conditions without the versatility to stay one jump ahead of its times which had characterized the first 55 years of its existence.

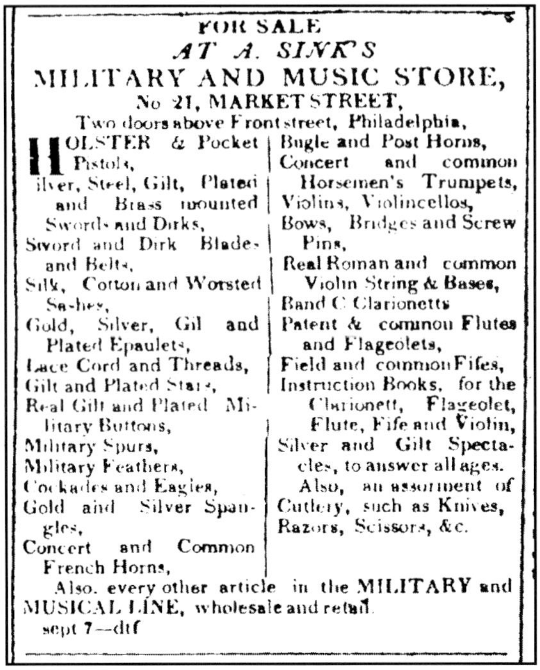

FOR SALE
AT A. SINK'S
MILITARY AND MUSIC STORE,
No 21, MARKET STREET,
Two doors above Front street, Philadelphia,

HOLSTER & Pocket Pistols,
Silver, Steel, Gilt, Plated and Brass mounted Swords and Dirks,
Sword and Dirk Blades and Belts,
Silk, Cotton and Worsted Sashes,
Gold, Silver, Gil and Plated Epaulets,
Lace Cord and Threads,
Gilt and Plated Stars,
Real Gilt and Plated Military Buttons,
Military Spurs,
Military Feathers,
Cockades and Eagles,
Gold and Silver Spangles,
Concert and Common French Horns,
Bugle and Post Horns,
Concert and common Horsemen's Trumpets,
Violins, Violincellos,
Bows, Bridges and Screw Pins,
Real Roman and common Violin String & Bases,
Band C Clarionetts
Patent & common Flutes and Flageolets,
Field and common Fifes,
Instruction Books, for the Clarionett, Flageolet, Flute, Fife and Violin,
Silver and Gilt Spectacles, to answer all ages.
Also, an assortment of Cutlery, such as Knives, Razors, Scissors, &c.

Also, every other article in the MILITARY and MUSICAL LINE, wholesale and retail.
sept 7—dtf

Advertisement of Abraham Sink, Horstmann's competitor, from the Philadelphia Gazette and Daily Advertiser, *January 7, 1825*

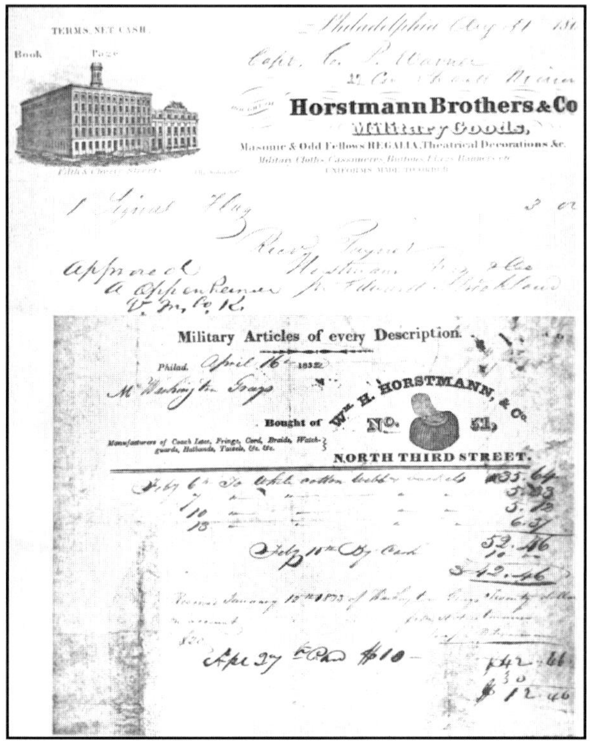

W. H. Horstmann & Co to the Washington Grays of Philadelphia. April 16, 1832. Washington Grays archives

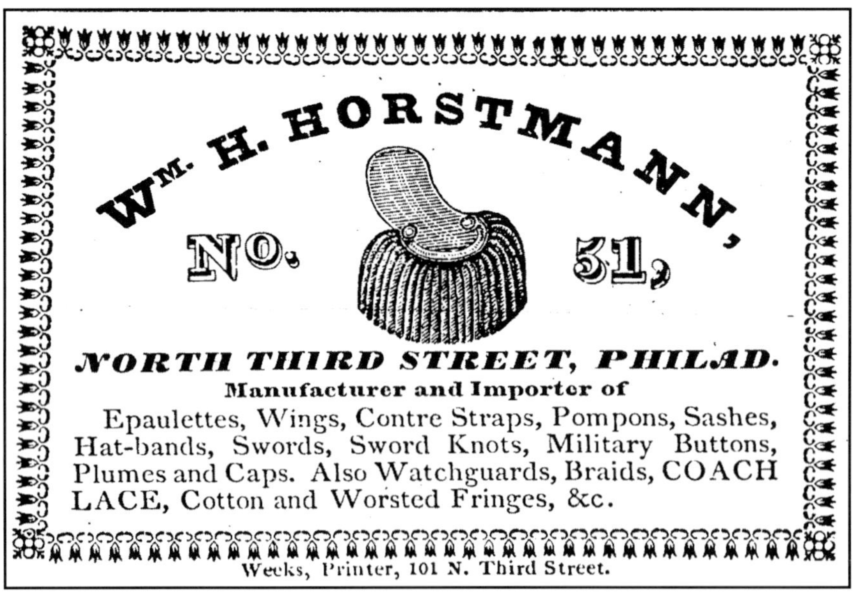

W^M. H. HORSTMANN,
NO. 51,
NORTH THIRD STREET, PHILAD.
Manufacturer and Importer of
Epaulettes, Wings, Contre Straps, Pompons, Sashes, Hat-bands, Swords, Sword Knots, Military Buttons, Plumes and Caps. Also Watchguards, Braids, COACH LACE, Cotton and Worsted Fringes, &c.
Weeks, Printer, 101 N. Third Street.

Horstmann label, 1830s. Horstmann scrapbook

Chapter 6. The Twentieth Century

The Horstmann firm had always consisted of two companies, one manufacturing, importing and selling civilian trimmings and the other engaged in equivalent activities dealing in a large variety of military accouterments. The firm had not engaged in the manufacture of garments, yet the tailoring of a variety of uniforms was very much what the trade as a whole demanded. The duality of a civilian and military business had existed from at least 1828 and had been expressed since the 1850s through the organization of two firms within the same general business. These firms were William H. Horstmann & Sons, which supposedly handled the civilian side of the trade but whose name appears on many Civil War military items, and Horstmann Brothers & Company, formed in 1858 to take on the military side of the business. While these two companies had slightly different management structures over the years, their reason for being separate was not very great, especially after the break-up of their New York company, Horstmann Brothers & Allien, in 1876. This new organization left Horstmann with only a sales branch in New York rather than an active partnership with the Alliens.

In Philadelphia, Horstmanns had remained a simple family partnership since 1845, but by the late 19th century business in general was tending more and more toward the incorporation of large firms like that of the Horstmanns. Over the years older, trusted employees had assumed places in the management structure, but Horstmann remained a family business. This "old guard" was passing from the scene during the 1880s with the retirement of William Schultz, Jr., for health reasons in January, 1885, and with the deaths of Albert Weihenmayer in February, 1887, and J. Howard Mecke in October of the same year. Accordingly it was decided in 1893 to consolidate and incorporate the company, to take effect January 1, 1894. The new firm which replaced William H. Horstmann & Sons and Horstmann Brothers & Co. was entitled "The William H. Horstmann Company".

The new incorporation did not stop the change in management. Samuel Clarkson, William J. Horstmann's brother-in-law, became the first president of the new firm only to die in August, 1894, after a few months as president. F. Oden Horstmann (1846-1894), the senior brother of William J. Horstmann's two male heirs, left the firm in 1893 and died in October 19, 1894.[128] This left Walter Horstmann (1857-1922) and his nephew, F. Oden Horstmann, Jr. (1877-1912), to represent the family management of the firm.

Walter Horstmann was a graduate of the University of Pennsylvania, class of 1877. *One Hundred Years* credits his active service with the company to 1876.[129] He succeeded to the presidency of the Horstmann firm on Clarkson's death in 1894. In addition to his role with the William H. Horstmann Company, Walter Horstmann was a member of Philadelphia's Chamber of Commerce, a director of the Philadelphia Bourse, and a trustee of a number of libraries and hospitals, all fitting his role in society. He was also on the vestry of St. Luke's Protestant Episcopal Church, the family having left their Lutheran origins.[130]

Oden Horstmann's son, F. Oden Horstmann, Jr., was apparently somewhat different as a manager. He died in Washington, from heart failure on June 30, 1912. His obituary in the *Philadelphia Evening Bulletin* the next day describes him as an expert golfer and as having sustained an injury on the Chevy Chase course a few weeks earlier which was held as an indirect cause of his demise. He graduated from Harvard in 190 and for a time worked in the family business but, like his father, moved to Washington where President Taft appointed him inspector of saloons in 1911. The contents of his apartment, including furnishings and trophies, was valued at over $50,000, a large figure for the time and a matter remarked upon in his obituary.

---

[128] *One Hundred Years*, pp. 28-9. Also, gravestone inscriptions in Laurel Hill Cemetery, Philadelphia.

[129] *One Hundred Years*, p. 34.

[130] *Encyclopedia of Pennsylvania Biography* Volume XIV (New York: Lewis Publishing Company, 1923), pp.271-272.

Prior to the reorganization, the Horstmann companies are best described in *Manufacturies and Manufacturers of Pennsylvania.* This article describes the factory at Fifth and Cherry which contained the firm's manufacturing areas, salesrooms, and offices in a six story structure 140 feet on Fifth Street and 200 feet on Cherry Street reaching back to Race Street. The business was divided into some thirty departments.

> "... More than 1000 distinct looms and machines are in use in the building, many very costly and some invented and exclusively used by this firm, the motive power being supplied by a steam engine of fifty horse power. The area covered by the works is about 14,000 square feet. The number of hands employed is about 500. ... The goods produced by this house are of almost endless extent and variety. They include goods woven from all the various textile fibers–cotton, wool, silk, etc.–in every color and pattern, and are used for an infinite number of purposes. Narrow woven goods are in staple production, made up into material for dresses of both sexes, for use in daily life, and for regalia for 'societies;' for the costumes of the stage, the upholstering of houses and of carriages, the uniforms of soldiers, together with equipments for the same, and for funeral purposes. The raw material used in the manufacture is to a great extent very costly, and their storerooms often hold as much as $200,000 worth of goods in an unworked state. There are two rooms devoted to power looms in the factory, one for coach lace and one for other styles of weaving, in which about 250 of these machines are constantly running. The braiding machines in the coach lace room are very noteworthy. ... The one article of swords may be taken as an instance. This trade grew naturally and immediately out of the established army and navy goods department of the works, it being necessary that the sword itself should be furnished with the sword-belt and other trappings all complete. Every part of the sword and trappings, with the exception of the blade, is made on the premises. The blades are almost all imported from the ancient German sword-blade emporium of Solingen. ... There is in this department a stock of some thousands of blades, of many different patterns and sizes, ready to be set and finished. Any style of sword can be had from this warehouse, from the plainest kind up to a presentation sword worth a thousand dollars. In addition to their own vast capacity

for manufacturing goods, this firm also import and sell large quantities of goods manufactured elsewhere. The house also has two stores in New York and an agency in Paris. . . ."[131]

The files of Linden Hall, the female counterpart of the Moravian affiliated school the Horstmann brothers had attended in the 1830s in Lititz, Pennsylvania, contain an interesting series of billheads which show Horstmann's trade evolution from 1868-1881. These bills for trimmings show William H. Horstmann & Sons as "Manufacturers and Importers of Ladies Dress Trimmings, Small Wares . . . Carriage Laces, Window Blinds & Upholsterers Trimmings". "Military Goods" are also noted as part of the stock of this ostensibly civilian oriented trade (June 27, 1868). The military occupation does, however, disappear by October, 1868, and remains missing on the billheads through 1871. The last billhead (February 11, 1881) notes the firm's dealings to include undertakers' trimmings, military goods, and society goods. While these omissions may only reflect the whim of whoever ordered the billheads, it is more likely that it reflects the lessened interest in military goods after the close of the Civil War contrasted with a heightened interest in such manufactures during the last twenty years of the 19th century.

The reorganized company of 1894 was described in detail within the centennial history of the firm written in 1916. This source notes a manufacturing department, consolidated as "Section D", which served two sales divisions. "Section A", the old William H. Horstmann & Sons, made Columbia yarns, knit goods, cottons, imported small wares, embroidery materials, art needlework upholstery, and limousine trimmings. This section dealt with the civilian side of the business. Its trademark was the personification of Columbia as shepherdess seated in a rural scene. "Section B," formerly Horstmann Brothers & Company, manufactured and sold military equipments and uniforms, society goods, trimmings, flags, banners, church, and costumers' materials. Its trademark was a knight in armor on a pedestal with a band around the figure reminiscent of the Order of the Garter but saying "Wm. H. Horstmann Company. Philadelphia".[132] In 1916, under President Walter Horstmann were vice president Samuel Eckert, who joined the firm in 1872; George Eiler, Jr., general manager who came to the firm in 1869; Harry Freund, treasurer, who had been with the business since 1856; Harry McManus, sec-

[131] *Manufacturies and Manufacturers of Pennsylvania of the Nineteenth Century* (Philadelphia: Galaxy Publishing Company, 1875) pp. 406-408.

[132] *One Hundred Years,* p. 34.

retary, who was hired in 1871; and directors Edwin S. Dixon (1894) and Sidney Sall of Toronto who came in 1908.

A surviving inventory of Horstmann button dies held by the Scovill Manufacturing Company dated 1903 is found among the Scovill papers at the Baker Library. This shows the range of Horstmann's business at the opening of the 20th century. In it are buttons for the uniforms of such diverse entities as the A & C Air Line Railroad, the Pennsylvania Railroad, the Los Angeles Fire Department, the Minneapolis Park Police, the U. S. Marine Corps made with Starkey back dies, the State Fencibles (Philadelphia militia), the "Ex Confederate" (Confederate Veterans), and a host of other military schools, U. S. and state militia military units, the old 1876 Centennial guards, and fire departments. While this shows a national scope of business, it all relates to sales of uniforms.

A number of Horstmann catalogs dating from the decade between 1910 and 1920 have survived. These show that the company had now conformed to the dictates of the trade by including uniforms prominently among such traditional Horstmann trimmings as epaulettes, aiguilettes, shoulder knots, dress belts, sabers, sashes, and leggins. The 1914 catalog addressed to Army officers includes these items but also includes camp gear such as canteens, mess kits, dispatch cases, blankets, binoculars, bed rolls, pistols, watches, trunks, horse equipments, and leggings. The uniforms included officers' dress blue and white garments along with civilian evening dress. Appended to the catalog is the advertisement, "Our facilities are exceptional for producing the very highest grades of uniforms and equipments."

These catalogs show Horstmann branching out in the same manner as their competition to have a military department store-like approach to completely outfitting the officers of the Army and Navy who were among their chief customers. Although the Horstmann stores had always sold more than they produced, the items now being handled represent a significant departure from their usual trade and a complete reversal of the founder's reliance on trimmings as the firm's primary line. While Horstmann may have been meeting its competition by expanding its line, it had ceased to function as the bold and decisive business leader which had characterized the firm in its first 75 years.

What was true of the military business can also be said for the civilian trade. Fashion had changed. The amount of ribbon required for female personal adornment had diminished, as had the need for lamp shade tassels and other staple Horstmann products. The manufacture of yarn and thread replaced them, complete with books giving instruction as to the proper use of Columbia yarns for sweaters, bootees, slipovers, and rugs. These utilitarian books issued by Horstmann are known as the *Columbia Book of Yarns*, apparently issued yearly from about 1900 till after 1936. These books put heavy emphasis on Horstmann's yarn trade and what could be knitted from these products. Gone are the shade tassels and coach lace trimmings of the middle years of the 19th century. As the trade changed, Horstmann attempted to keep up with the new business realities. Until at least 1922, the year of Walter Horstmann's death, this effort appears to have been successful.

The business employed over 600 people in 1916, 200 of them in clerical and sales departments.[133] Branches were in New York at 4th Ave, and 18th Street and Baltimore. The Philadelphia plant covered about four acres. Three employees had been employed there for over 50 years, 21 from 30-40 years; 35 from 20-30 years. Thus 69 persons had been with the Horstmann company 20 years or more in 1916.[134] The company had employee benefits such as a relief organization which provided sick and death benefits, a recreation room for social activities and concerts given by the YWCA, a restaurant, a safety association, athletic association, bulletin board, and even gifts of a membership in the Franklin Institute for deserving employees. The company had always been interested in safety, and the building was originally designed with broad staircases to which was now added an automatic sprinkling system. The company newspaper "Chat" apparently contained articles relating to work as well as items of "educational value."[135] Unfortunately, no copy of this paper has yet been located.

Horstmann had a purchasing branch in Paris, manufacturing plant in Lyons and selling agencies in New York, Boston, and Baltimore as well as ". . . a force of travelers covering the whole country." It describes its own business as excellent with continued expansion of territory and demand. It had already issued over 80 catalogs, of which 40 were for "secret societies." The civilian business published instruction books on knitting, needlework, embroidery and ". . . the well known Columbia Book of Yarns. . . ."[136]

The factory complex or "mill" occupied a section on the fifth floor 60 x 100 feet devoted to the manufacture of cords and chenille. There

[133] *One Hundred Years*, pp. 34-37.

[134] *Ibid.*, pp. 36-37.

[135] *Ibid.*, pp. 38-39.

[136] *Ibid.*

were areas on the fifth floor devoted to hand worked upholstery trimmings and fringes, and another department for skeining, boxing, labeling, and packing of yarns and threads, and another for braiding machines and the manufacture of flags by power driven sewing machines. On the third floor was silk processing consisting of a "throwing" operation where raw silk from China was soaked, separated, and spooled. Also on the third floor were coach lace looms. The second floor was for power loom weaving with over 40 large looms, some 25 feet in length with Jacquard attachments. These looms had multiple banks of shuttles and turned out upholstery, military, and church trimmings, including the silk hat bands embroidered with ships' names used by the U. S. Navy. On the first floor "fancy goods" were manufactured with patterns cut and fancy work laid out. A whole battery of power sewing machines turned out lingerie, infant's wear, cushions, doilies, luncheon sets, table covers, and pillow cases. Also on the first floor was a stock room while the basement was for twisting and spooling "Columbia yarns."

The manufacturing departments of "Section B" were related to the military and society side of the business and were somewhat smaller. The metal shop was for brazing, polishing, and plating. Sword guards were cast, ground, polished, and plated and blades fitted with care. Scabbards, fittings, and belts were also manufactured. *One Hundred Years* cites 101 metal parts stored in racks, ready for hand and machine operation. Woodworking was conducted in another shop. Flag staffs, lodge furniture, and poles were made in this area.

Another machine shop actually made the repairs to existing machines and built or made changes to other machines. Significantly, the description of the 1916 plant includes a separate building which housed the "Uniform Department" where cloth was turned into finished garments. Another building housed a leather department for manufacturing leggings, leg protectors for foundry workers, service belts, horse equipment including saddles, bridles, and guards, sword scabbards and cases, single unit chapeaux, epaulette and belt cases, suitcases, ". . . bugle straps, flag belts, music pouches, helmets and hats for military, firemen and secret society use. . . ." Another room manufactured regalia consisting of all sorts of robes and garments and embroidery done entirely by hand. The rest of the complex contained a packing department, a "counting room" for handling the fiscal workings of the plant and various sales rooms, ware rooms and offices.[137]

All things change. Seven years after the centennial celebration of

[137] *Ibid.*, pp. 34-59.

the company in 1916, Walter Horstmann died on December 30, 1922. He left two daughters. His brother, F. Oden Horstmann, Sr., also left two daughters. His son, F. Oden Horstmann, Jr., died childless. The heirs of Sigmund Horstmann had been out of the picture since the retirement and death of William H. Horstmann (II) in the 1870s. According to the values of the time, the company, which had always been run by the male heirs to the business, was now out of Horstmanns.

The Urban Archives at Temple University hold the clippings files compiled by the *Philadelphia Bulletin*. These files are drawn from all Philadelphia newspapers and have proven invaluable in filling in pieces of the rest of the story. The *Public Ledger* of July 9, 1924, carried the lead article, "Group of Employees to Buy Firm of W.H. Horstmann Co". This article proceeds to describe a statement by George F. Schaefer, president of the company, stating that it had been decided at a meeting of the stockholders, to sell the company to ". . . a large group of its employees. . . ." The officials of the company, including Schaefer himself who had started as a pay-roll clerk 32 years previously, had risen through the ranks and they, with other employees, were buying the company from the Horstmann estate for $1,075,000. The old company had paid 7% dividends for 30 years and the enterprise was seen as sound and profitable. A new corporation, capitalized at $1,000,000, was formed for the purpose. More than 50 employees had been with the company 25 years and stock was offered on the basis of efficiency and loyalty. The new firm would concentrate on the production of yarns, knit goods, stamped goods, and art needle work.[138]

The *Public Ledger* article's allusion to the new corporation carrying on the civilian side of the business was correct. The military section of the business was evidently sold quite soon after to the Lilley-Ames Company of Columbus, Ohio–a traditional Horstmann competitor. Lilley ran the Horstmann company quietly, keeping the Horstmann name through the end of World War II and not emphasizing the change in ownership.[139]

---

[138] Temple University. Urban Archives. Philadelphia Bulletin Clipping File. This file contains several articles from various Philadelphia newspapers documenting the sale of the Horstmann company to its employees in 1923. The corresponding sale of the military branch of the company to Lilley-Ames is only documented through the Horstmann Uniform Company bankruptcy files now in the National Archives, Philadelphia. District Court of the United States for the Eastern Region of Pennsylvania. Docket 22693.

[139] National Archives, Philadelphia Branch, District Court of the United States for the Eastern District of Pennsylvania. Bankruptcy Docket 22693. 1947 Creditors Meeting of April 20, 1948.

In his bankruptcy deposition, Richard Wood, formerly the treasurer of the company, states that Lilley Ames took over Horstmann, he believed about 1921, and operated it as a division of Lilley-Ames until December 1945 when it was sold to a group of officers of Lilley-Ames. The company received the name "The Horstmann Uniform Company Successors to Uniform and Regalia Business of William H. Horstmann Company." In this configuration it weathered the depression and changing circumstances. These included the lessened interest in fraternal and military organizations and the ability of officers to purchase their uniforms, accessories, and gear at the Post Exchange, which made the large military goods firms obsolete.

In the 1920s and '30s the two companies, William H. Horstmann Co. and The Horstmann Uniform Company, ran with separate addresses but in the same building, Fifth and Cherry. The civilian and military businesses are listed in the Philadelphia City Directories and phone books through July, 1939, at 526 Cherry and Fifth and Cherry respectively, but the William H. Horstmann Company is not listed in April, 1940, or thereafter. A search of the Bankruptcy Index for those years does not indicate any action which closed the doors of the William H. Horstmann Company. It is probable that the company simply sold out or quietly closed leaving no successor.[140]

In December 1945 Lilley-Ames incorporated the Horstmann Uniform Company in Pennsylvania, and then sold all the stock to the group of former employees who purchased Horstmann for $50,000, some of which was apparently never paid. The company manufactured uniforms for the military, police, bands, and even some civilian clothing.[141] The new owners took out loans on real estate, goods, and equipment in spite of which they still owed tax monies to the Federal government. There were several efforts to reorganize, but to no avail, and the stock of the defunct company, including the name, was sold June 29, 1948, at public auction. The auction sale included the contents of the entire Fifth and Cherry Streets building, which by that time was mostly garments, hats, sewing machines, and other miscellaneous appurtenances of a uniform manufacturer. One lot of "paper" which, according to Duncan Campbell, included the company's records was sold to a Mr. Mossman for $37.50. It was a sad end to a great venture.

---

[140] National Archives. Philadelphia Branch. Bankruptcy Index Clerk U. S. District Court Volume 5. May, 1937-December 3, 1944.

[141] National Archives. Philadelphia Branch. Bankruptcy files of the District Court of United States for the Eastern District of Pennsylvania. Docket 22693.

At the same time, the bankruptcy proceedings allow us to examine the Horstmann Uniform Company in its last days. Its actual mechanics, while protracted, are fairly simple. The purchasers of the firm from Lilley-Ames found the trade in uniforms difficult after the close of World War II. They had issued stock in the amounts of $24,000 common and $26,000 preferred stock–about 2000 shares. This nominally capitalized the company at $50,000. However, in the testimony of Richard Wood at the creditors' meeting of April 20, 1948, it was stated that in fact the stock subscriptions were assumed by the officers, who were to have paid for the stock through salary deductions. Because ". . . half the time we never drew salaries . . ." the stock remained unsubscribed and the company uncapitalized.

With the stock never fully paid for by the company officers, its sole purpose was to retain control of the company. When the funds actually deposited proved insufficient, the officers mortgaged the company to an entity called the Manufacturers Trading Corp. of Cleveland in notes dated August 31 and September 3, 1946. They probably had the only real money in the new company. The real estate was valued at $43,300 and the inventory/accounts receivable at $35,196.31, on which $22,000 and $37,620 were claimed as due by Manufacturers Trading Corp. at the time of bankruptcy in October, 1947. The Manufacturers Trading Corp. utilized the Lawrence Warehouse Company to watch over the inventory used to secure their mortgage. Lawrence Warehouse seems to have segregated part of the firm's inventory as security and locked it away. This stock could be recovered if the Horstmann Uniform Company paid 50% of its cost price to Manufacturers Trading. As cash receipts came in, they were deposited and used to pay off the note to Manufacturers Trading Corp.

The bankruptcy proceedings began in May 1947. After attempts to reorganize failed, the claim for Federal taxes was filed February 12, 1948. In addition to Federal taxes, there were many secured and unsecured creditors whose claims could not be paid and the firm's assets were ultimately sold at auction on June 29 of that year. A final order of distribution was filed March 23, 1949.

The firm's financial statement, filed on August 31, 1947, showed total assets of $88,024.78 and liabilities of $104,351.23, not an impossible amount to make up. In fact, in the final dissolution of March, 1949, creditors got 67.1% of their money. What had changed was the nature of the trade. The "secret societies" while still active, were on the decline. Officers now purchased their uniforms, for the most part, through the Post

Exchange system and not through large military goods firms like Horstmann. Horstmann's great competitors such as Lilley-Ames, Horstmann's former owner; H. V. Allien & Co., Horstmann's former New York branch, and a host of other suppliers would soon join Horstmann in bankruptcy. Only Pettibone Uniform Company of Cincinnati and Jacob Reed & Sons of Philadelphia survived long enough to see the 1990s, but now they too are gone. The new contracting systems in the military and the changing emphasis of the times had made them into dinosaurs.

Among the Horstmann bankruptcy schedules are the lists of creditors (schedule A-3) listing those firms upon which Horstmann relied for its business supplies. Among them are GEMSCO, a New York insignia conglomerate to which Horstmann owed $43.65; V. H. Blackinton & Co of Attleboro, Mass., another insignia manufacturer, which was owed $110.67; H. D. Folsom Arms Co. of New York $39.91; J. R. Gaunt & Son, an English military house whose New York branch was owned at that time by Waterbury Button Company, $84.62. Lilley-Ames was owed two accounts, $356.27 and $19.21, and J. H. Ferd. Hahn, another insignia producer, $424.42. The *Howitzer* and *Lucky Bag*, the West Point and Annapolis yearbooks, were owed $225 and $80 respectively for advertising Horstmann's uniforms. Waterbury Companies, the successors to Waterbury Button Company, was owed $109.27, certainly for uniform buttons. The Woolen Corp. Of America had a credit of $409.63.

The listings of Horstmann's stock sold on June 29, 1948, and other appraisals done for the auction reflect the nature of its business. Included were such items as 75 signal flags, 361 pair insignia, 685 yds. fringe, 28 belts and holsters, spurs, 218 uniforms 105 pair white gloves, 340 overseas caps, 420 pair infantry insignia, gold and silver braid, 117 boxes uniform buttons, woolen trousers and breeches, lining (for uniforms), and uniform jackets. Another list shows 5 Knights Templar uniforms, 231 Army officers' caps, 23 band uniforms, 5 police uniforms, 216 Army officers' uniforms, 49 Sam Browne belts, 430 gross buttons, boxes of uniform trimmings, 20 Stetson hats, uniform shirts, 186 uniform hat frames, pompons and tassels, U. S. flags, 10 swords, patches and chevrons, graduation robes, wool, cotton and chino cloth, unfinished jackets and linings, buckles, thread, clasps, and remnants.[142]

The appraisal of machinery dated January 1, 1947, shows such items as winding and spooling machines, several 16, 20, 28, and 48 spindle

[142] National Archives, Philadelphia Branch. Bankruptcy files of the District Court of the United States for the Eastern District of Pennsylvania. No 22693.

braiding machines, a knitting machine, a Lewis blind stitcher, 12 Singer sewing machines, 10 embroidery machines, 2 winding machines, and a twisting machine.[143]

These examples of stock and machinery are far removed from the 19th century inventories of the Horstmann firm. Not only are they much smaller in value but they represent a very different trade configuration. While still subcontracting such items as buttons and insignia, the Horstmann Uniform Company was in fact very different from its predecessors. The machinery and stock reflect a standard, middle sized uniform manufacturer, one of a large number of such firms throughout the United States. In its attempts to conform to its trade and thus survive, the Horstmann firm had been effectively gutted of its traditional trade in embroidery, military laces, sword manufacture, accouterments–those operations which had distinguished the 19th century firm. Horstmann had run out of trade options. It was time for Horstmann to close its doors.

---

[143] *Ibid.*

Chapter 7.
Conclusion

As this work was initially envisioned, the Horstmann companies were to serve as a vehicle by which we can understand both the military goods trade as a whole and the story of the many other firms whose records probably have not survived. But as Horstmanns' history unfolded, it became evident that the Horstmann longevity was no accident. Because of the firm's importance, some of its records were thought worthy of preservation and enough has survived to permit this study. Horstmann was unique, not just in its survival for one hundred and thirty odd years, but in its sequence of talented managers through three generations of one family.

The history of the Horstmann business is in many ways the story of the family of its founder. While they ran the business, it remained profitable. When they died out, the firm was unable to replace them with similar leadership. Given that times were becoming more difficult for traditional military goods and trimmings firms after World War I, Horstmann would have had to undertake some bold realignment in a new field in order to survive. Instead, with the Horstmann family now out of the picture, the company chose to try to hang on to their yarn business and to sell off the remainder of their traditional line, namely the entire military trade.

Because of Horstmanns' longevity, the business participated in much of the economic development in the United States from 1800-1950. In its first days, William H. Horstmann competed with both craftsmen and import merchants. Only a few of these early firms attempted to both manufacture and import the goods they sold. Mostly they were either small manufacturers or import merchants competing for the retail market. Horstmann was able to utilize both his craft and factory management background in Europe to initially set up his business in 1816, then, from at least the early 1820s, to import goods which actually competed with his manufacturing business. By 1828, Horstmann was both manufacturing and importing a full line of merchandise in both his military and civilian business enterprises.

Horstmann's business successes enabled his firm to survive into the next phase of competition, that of the capitalist mill owner. W. H. Horstmann had already proven himself ahead of his time by importing advanced looms and other machinery from Europe within ten years of founding his business. By the 1830s, Philadelphia was developing a large number of textile mills in Kensington and Roxborough. Here, W. H. Horstmann's two sons proved their abilities to manage a large and physically diverse business and to invent advanced machinery to better serve the manufacturing processes in Horstmann's center city business and at their new factory in Kensington. Finally, the brothers were able to envision expanding their business beyond its traditional manufacturing reliance on woven goods and to assemble their new diverse manufacturing functions at Horstmanns' state-of-the-art factory at Fifth and Cherry Streets. Because of their business skills, the brothers had outdistanced their smaller rivals like Henry Korn, Bartholomew Delapierre, William Pinchin, and Peter Mintzer. They were now in competition with the largest military and silk weaving manufacturing businesses such as the Ames Manufacturing Company of Chicopee, Massachusetts, in the arms trade or Henry Cressman, who was conducting a national business in military headgear from Philadelphia. They also competed with large retailing and importing houses like those of Francis Tomes, Young Smith & Co. or Schuyler Hartley & Graham, all of New York. Horstmann's advantage lay in the firm's ability to combine the functions of manufacturer, importer, and retailer in a way their competition never attempted.

After the Civil War, the largest military goods businesses were not only military department stores, like Horstmann, but were dependent upon the tailoring of uniforms as a mainstay, or they simply moved completely into the areas of dealing in arms. Horstmann, then managed by the third generation of the founder's family, had managed to implement a thriving trade in swords while giving up other experiments such as manufacturing drums and insignia. However, Horstmann had never been a tailoring house and only slowly moved to meet this aspect of its competition, while still managing to neatly combine manufacturing with import in other aspects of their trade. By the late 19th century, their competition also had developed extensive manufacturing capabilities, usually based upon the manufacture of uniforms. Such uniform making companies as M. C. Lilley, Pettibone, Frank Henderson, Ridabock, and George Evans never needed to participate extensively in the import trade, which was one of Horstmann's traditional strengths. However, Horstmann eventually moved to meet the competition and began to manufacture uniforms. The comparative rarity of surviving uniforms with the Horstmann label probably indicates that the firm was never as reliant upon this part of the trade as was their competition, at least until the very last days of the firm.

Horstmann remained strong through all of its incarnations during the 19th and early 20th centuries. Changing business conditions and periodic economic downturns seemed well within its abilities to cope. However, this situation could not last forever. In fact, the conservative business practices which had served the Horstmanns well before 1916 became liabilities after World War I. As the 20th century progressed, times were no longer on Horstmann's side. The ability of the company to keep ahead of business and economic conditions had changed even before 1900, and the firm was just one of the pack. New innovations were no longer as welcome as they had been under the original Horstmann Brothers. Those Horstmanns had extended their father's business by expanding their product line and appealing to a wider customer base.

The newer management was content to continue to produce the products which had made their fortune. This was acceptable so long as times were good and there was a demand for Horstmann products, but the two firms bearing the Horstmann name after 1923 were in no condition to take on both the Depression and the changing military goods business after World War II.

The new business situation was one where military officers normally purchased their uniforms and equipment through the Post Exchange and the fraternal business was declining, creating a reliance upon police and band customers. This was not enough to carry the Horstmann company even if it had been well managed.

So Horstmann became a dinosaur and closed its doors. Its factory complex was torn down a few years after the firm closed to make way for the new United States Mint. However, Horstmann's legacy lives on through the thousands of artifacts it produced in its years of prosperity,

They maintain an interest in Horstmann which would not be there if the company sold, say, cut nails. Given this interest, Horstmann will be with us for many years. Luckily, the size and prosperity which the firm attained created a situation in which family members deemed it important that something of the firm survive. Through these historical sources we can understand not only Horstmann but something of the business world as America grew up.

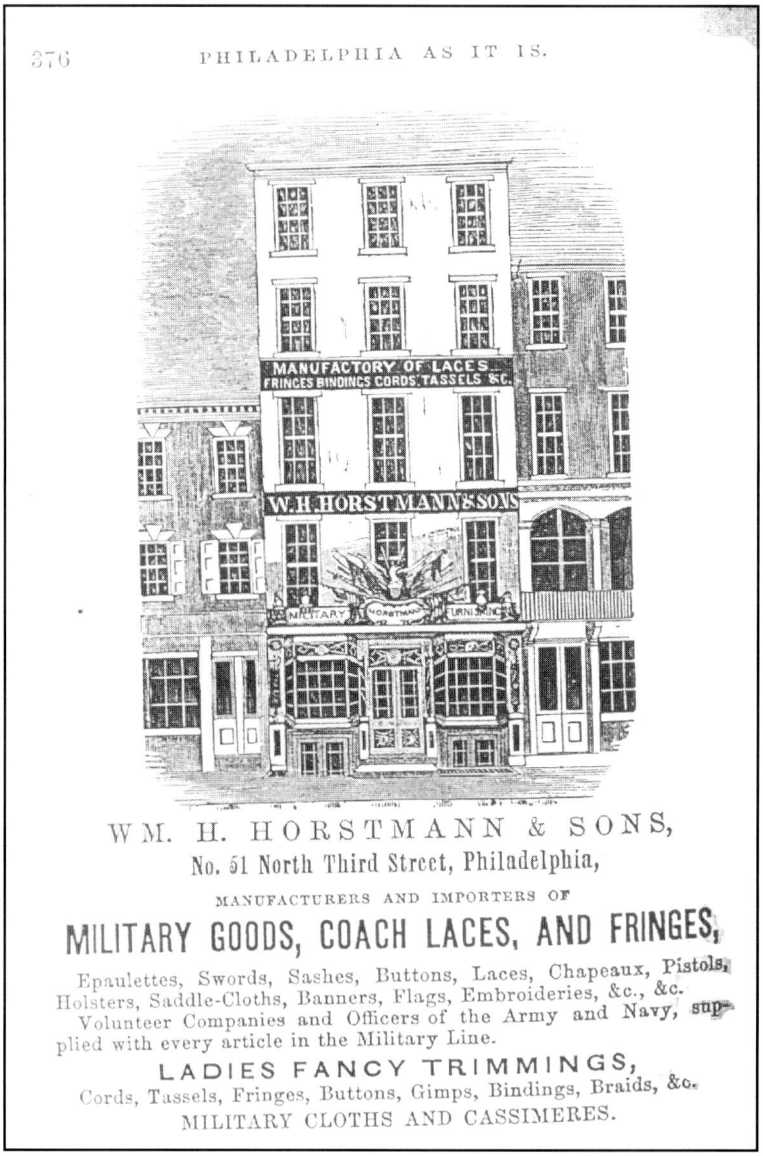

Horstmann advertisement from A. R. Smith's Philadelphia As It Is *in 1852*

Bibliography

Ames Sword Company: *Illustrated Catalogue and Price List of the Ames Sword Company, Chicopee, Mass. U.S.A.*, c. 1882.

Bazelon, Bruce S.: *Swords from Public Collections in the Commonwealth of Pennsylvania* (Lincoln, R.I.: Man at Arms, 1985)

Bazelon, Bruce S. & McGuinn, William: *A Directory of American Military Goods Dealers and Makers* (Manassas, Va: REF Publications, 1995)

Beck, John: *Complete Catalogue of the Names of All the Students Who Have Attended the Lititz Academy for Boys, from A.D. 1815 to A.D. 1865* (Lancaster: Pearsol & Geist, 1865)

Biographical Encyclopedia of Pennsylvania of the Nineteenth Century (Philadelphia: Galaxy Publishing Company, 1974)

Brown, M.L.: *Firearms in Colonial America* (D.C.: Smithsonian Press, 1980)

The Cambridge Economic History of Europe (Cambridge: Cambridge Union Press, 1977) 8 volumes

Cameron, Rando: "French Gilds Under the Old Regime" published in *Essays in French Economic History* (American Economic Association, 1970)

Catalogue Regulation Uniforms & Equipments for Officers of the U.S. Army (Philadelphia: William H. Horstmann Company, 1914)

Collections of the Genealogical Society of Pennsylvania (Philadelphia, 1896) Volumes 8, Volume 64 (1901), Volume 80 (1904)

The Columbia Book of Yarns (Philadelphia: Wm. H. Horstmann & Co, 1920)

Connelly, Owen: *Napoleon's Satellite Kingdoms* (New York: Free Press, 1965)

Cutts, Mary Peperell Sparhawk: *The Life and Times of William Jarvis of Weathersfield, Vermont* (New York: Hurd and Houghton, 1869)

Doerflinger, Thomas N.: *A Vigorous Spirit of Enterprise: Merchants and Economic Development In Revolutionary Philadelphia* (Chapel Hill: University of North Carolina Press. Institute of Early American History and Culture, Williamsburg, 1986)

Duy and Woods Insurance Company Map, 1872. Print Collections, Historical Society of Pennsylvania Of603P.

Filby, P. William: *Philadelphia Naturalization Records* (Detroit: Gale Research Company, 1982)

Freedley, Edwin T.: *Philadelphia and its manufactures: A Hand-Book* (Philadelphia, Edward Young, 1858 and 1859 editions)

Garagliano, Leonard J.: *Abstract, Review and Notes Regarding Principal Sword Contractors During the American Civil War* (Salisbury, MD: Peninsula Press, 1983)

J. W. H.: *Marcellus Hartley A Brief Memoir* (Privately Printed, 1903)

Hexamer, E. & Son. Philadelphia. *Hexamer General Surveys* Philadelphia Free Library Map Room. Maps; Horstmann's 1865, p. 9;1866, p. 81; 1872 pp. 626-627; 1887 pp. 2093-2094.

Heyl, Lewis: *United States Duties in Imports Published in 1874* (Washington: W. H. & O. H. Morrison, 1874)

Heitman, Francis B: *Historical Register and Dictionary of the United States Army* (Washington, Government Printing Office, 1903) 2 volumes

Horstmann Bros. & Co. Illustrated Catalogue of Military, Society Regalia, Theatrical Goods, Flags, Banners Etc. (Philadelphia: A. C. Bryson & Co, 1877)

Horstmann Brothers & Co. Price List G. A. R. Supplies 74 (Philadelphia: Horstmann Brothers & Co., ca. 1890)

Horstmann Company: *One Hundred Years 1816 to 1916, the Chronicles of an Old Business House in the City of Philadelphia* (Philadelphia: William H. Horstmann, 1916)

Illustrated Catalogue and Price List of the Ames Sword Company Chicopee, Mass. (Ames Sword Company, ca. 1880)

Leonard, John W (ed.): *Who's Who in Pennsylvania* (New York: L. R. Hammersley & Co., 1908)

Manufacturers and Manufactories of Pennsylvania in the Nineteenth Century (Philadelphia: Galaxy Publishing Company, 1875)

Margerand, J. "Les Coiffures de l'Armee Francaise" *Revue Mensuelle Illustrée* (1909-1924) 2 Volumes

McElroy, A: *Philadelphia City Directory for 1845* (Philadelphia: Edward C. & John Biddle, 1845)

McElroy, A.: *McElroy's Philadelphia City Directory for 1865* (Philadelphia: A. McElroy, 1865)

Mills, Anson: *My Story* (D.C. Byron Adams, 1921)

Morris, Charles: *Makers of Philadelphia* (Philadelphia: L. R. Hammersley and Company, 1894)

National Archives: *Lists of vessels arriving at Philadelphia, 1800-1882* Microfilm number 425

National Archives. Philadelphia Branch. *Passenger Index, Philadelphia 1800-1906* (Microfilm Reel M360)

Neville, R.E. Jr.: "James Conning, Confederate Sword Maker" *Bulletin of the American Society of Arms Collectors* Number Thirty-Eight. Spring 1978

Our Firemen: A History of the New York Fire Departments (New York: Augustine E. Costello, 1887)

Philadelphia Yearbook, 1917 (Philadelphia, Philadelphia Chamber of Commerce, 1917)

Rae, Julio H.: *Rae's Philadelphia Pictorial Directory & Panoramic Advertiser* (Philadelphia: Julio H. Rae, 1851)

Sauers, Richard A.: *Advance the Colors* (Harrisburg: Capitol Preservation Committee, 1987) 2 volumes

Savage, J.R.: *The Philadelphia Circulating Business Directory for 1838* (Philadelphia: Morris's Xylographic Press, 1838)

Schmucker, Anna: *The Columbia Book of Yarns* (Philadelphia: Wm. H. Horstmann & Co, 1918)

Scranton, Philip: *Proprietary Capitalism: The Textile Manufacture at Philadelphia 1800-1885* (Cambridge: Cambridge University Press, 1983)

Smith, Charles J.: *Memorial of the Great Central Fair for the U.S. Sanitary Commission.* (Philadelphia: 1864)

Smith, R.A.: *Philadelphia As It Is In 1852* (Philadelphia: Lindsay & Blakiston, 1852)

Spofford, Ernest (ed.): *Encyclopedia of Pennsylvania Biography* (New York: Lewis Publishing Company, 1923) Vol XIV, Vol XVII (1928)

Tibbits, George W: *A Brief Sketch of the Cleveland Greys Outlining A Few of the Principal Happenings During A Period of Sixty-Six Years* (Cleveland: Cleveland Greys, 1903)

Wainwright, Nicholas B: *Philadelphia in the Romantic Age of Lithography* (Philadelphia: Historical Society of Pennsylvania, 1958).

Appendix 1.
The Horstmanns, 1785-1922

Johann Heinrich Horstmann M Dorothea (unknown)

Johann George Horstmann Brother W.H.H. ??? Daughter Sophie Dorothee Horstmann b. 12/27/1815	Johann Wilhelm Heinrich Horstmann (William H. Horstmann) 8/16/1785-8/15/1850 Business: 1816-1845 M Sarah Hoeckley 10/16/1792-1/17/1856	Sophia Magdelaine B 2/24/1788 Essen Cassel

Children of William H and Sarah Hoeckley Horstmann

William J. Horstmann 10/7/1818-5/18/1872 M Emma Walker 8/16/49 6/20/1826-1/28/1893	Sigmund H. Horstmann 7/23/1821-3/29/1870 M Elizabeth Cowgill West 11/1/1822-11/2/1876	Henrietta G Horstmann 11/13/1817-2/11/1884	Josephine Horstmann D. 12/29/1838

Children of Wm J. Horstmann and Emma Walker Horstmann

Walter Horstmann	Wm Henry Horstmann	F. Oden Horstmann	Clara Louise Thomas
4/4/1857-12/30/1922	4/27/48-8/16/1849	1/4/1846-10/18/1894	9/14/1847-2/27/1874
President 1884-1922		Left business in 1893	M
M			Newton Thomas
Margaret Dixon			(Child)
7/8/1859-10/21/1951			
(Children)			

Grandson
F. Oden Horstmann Jr.
10/3/1877-6/29/1912
Director 1903-1912

Children of Sigmund H. Horstmann and Elizabeth West Horstmann

Sallie Horstmann	Elizabeth Horstmann Lippincott	William H. Horstmann (II)
12/24/1850-3-23/1870	C1854-12/23/17	C1856-1/1/1879
	M. 10/21/1879	Retired 5/18/1877
	Walter Lippincott	M. Sarah Buy(D 2/26/1960)
	(Children)	(Children)

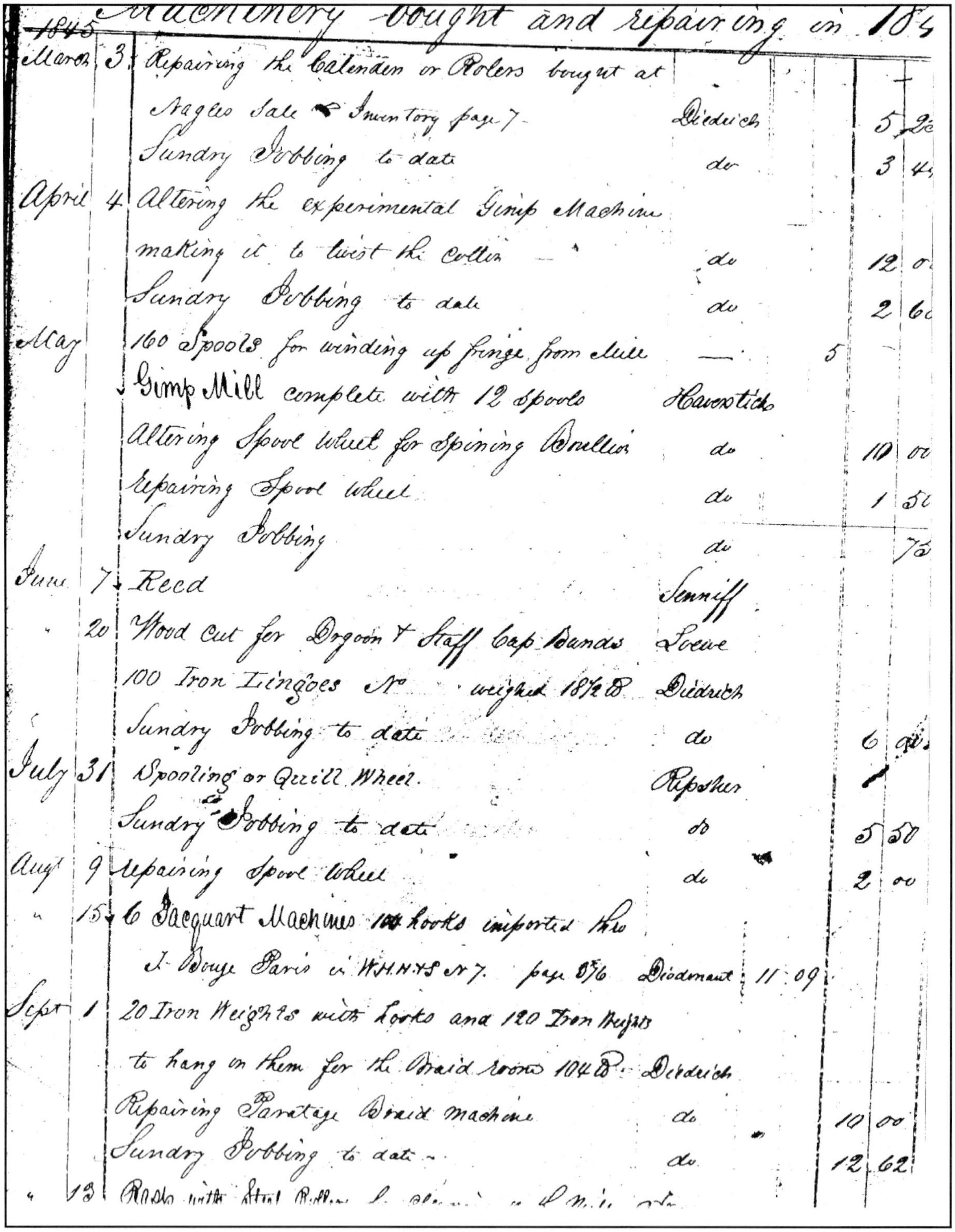

1845 Machinery bought and repairing in 1845

1845					
March	3	Repairing the Calender or Rolers bought at Nagles Sale & Inventory page 7	Diedrich		5	2[illegible]
		Sundry Jobbing to date	do		3	4[illegible]
April	4	Altering the experimental Gimp Machine making it to twist the Cotton	do		12	0[illegible]
		Sundry Jobbing to date	do		2	6[illegible]
May		160 Spools for winding up fringe from Mill	—	5		
		Gimp Mill complete with 12 Spools	Havenstick			
		Altering Spool Wheel for Spining Bullion	do		10	00
		Repairing Spool Wheel	do		1	50
		Sundry Jobbing	do			75
June	7	Reed	Senniff			
	20	Wood Cut for Drgoon & Staff Cap Bands	Loewe			
		100 Iron Lingoes No weighed 18½ lb	Diedrich			
		Sundry Jobbing to date	do		6	[illegible]
July	31	Spooling or Quill Wheel	Ripsher		[illegible]	
		Sundry Jobbing to date	do		5	50
Aug	9	Repairing Spool Wheel	do		2	00
"	15	6 Jacquart Machines 104 hooks imported thro J. Bouge Paris in W.H.H.&S No 7. page 8/6	Diodemant	11.09		
Sept	1	20 Iron Weights with hooks and 120 Iron Weights to hang on them for the Braid room 104 lb	Diedrich			
		Repairing Paratige Braid machine	do		10	00
		Sundry Jobbing to date	do		12	62
"	13	[illegible]				

Machine order entries March-September 1845. These transactions reflect the rapid growth and variety of Horstmann's trimmings business. Note the manufacture of dragoon and staff cap bands on June 20 and the Jacquard machines from Paris on August 15.

Appendix 2. Horstmann Addresses

PHILADELPHIA

1817	William Hortman (sic)	Fringe and Lace Weaver	12th & Arch
1818-24	William H. Horstmann & Co	Fringe & Lace Weavers	45 & 55 N 3rd
1825	William H. Horstmann	Coach Lace & Fringe Man.	53 N 3rd
1828-29	William H. Horstmann	Military Store	48 N 3rd
1830-42	William H. Horstmann	Military Store	51 N 3rd
1852-53	William J. Horstmann	Dress Trimmings	204 Chestnut
1843-54	William H. Horstmann & Sons	Military Store	51 N 3rd
1854	William H. Horstmann & Sons	Military Store/Trimmings	7th & Chestnut
1855-57	William H. Horstmann & Sons	Military Store/Trimmings	223 Chestnut
1858-59	William H. Horstmann & Sons	Military Store/Trimmings	723 Chestnut
1859-62	William H. Horstmann & Sons	Ladies Dress Trimmings	723 Chestnut
	Horstmann Bros. & Company	Military & Society Goods	723 Chestnut
1862-93	William H. Horstmann & Sons	Trimmings	5th & Cherry
	Horstmann Bros & Company	Military Goods	5th & Cherry
1893-1923	The William H. Horstmann Co	Trimmings Military Goods	5th & Cherry
1927-1940	The William H. Horstmann Co	Yarns & Knit Goods	423 Cherry
1927-1948	Horstmann Uniform Company	Military & Church Equip.	526 Cherry

HORSTMANN PERSONAL LISTINGS: PHILADELPHIA

1825	William H. Horstmann	91 N 3rd
1828	William H. Horstmann	48 N 3rd
1829-45	William H. Horstmann	68 Mulberry
1845-46	William J. Horstmann	166 N 5th
1847-48	William H. Horstmann	68 Mulberry

1847-8	William J. Horstmann	179 N 7th
1847-8	Sigmund H. Horstmann	239 N 6th
1849	William J. Horstmann	179 N 7th
1849	Sigmund H. Horstmann	68 Mulberry
1849	William H. Horstmann	251 Vine
1850-52	Sigmund H. Horstmann	68 Mulberry
1850-54	William J. Horstmann	191 N 5
1850-51	William H. Horstmann	251 Vine
1852-57	Sarah Horstmann	251 Vine
1853-57	Sigmund Horstmann	Marshall & Noble
1857-8	Sigmund Horstmann	251 Vine
1855-57	William J. Horstmann	249 Vine
1858-61	William J. Horstmann	705 Vine
1859-64	Sigmund Horstmann	707 Vine
1861-71	William J. Horstmann	1428 Arch
1872	William J. Horstmann	2106 Walnut
1874-79	Emma, Wid. William J.	2106 Walnut
1865-70	Sigmund H. Horstmann	1600 Walnut
1872-74	Elizabeth, Wid. Sigmund	1600 Walnut
1867-83	Henrietta Horstmann	4815-16 Germantown Ave.
1872-80	F. Oden Horstmann	2041 Arch
1869-71	F. Oden Horstmann	106 N19
1872-1882	F. Oden Horstmann	2041 Arch
1883-1885	F. Oden Horstmann	1346 Broad
1887	F. Oden Horstmann	3925 Chestnut
1893	F. Oden Horstmann	2126 Chestnut
1893	F. Oden Horstmann Jr.	2126 Chestnut
1894	F. Oden Horstmann	Washington D.C.
1879-1880	Walter Horstmann	2106 Walnut
1882-3	Walter Horstmann	4031 Walnut
1885-95	Walter Horstmann	238 N13
1896	Walter Horstmann	Elmwood Ave
1899-1903	Walter Horstmann	1206 Spruce
1921-23	Walter Horstmann	1804 Delancey
1935-1940	Mrs Walter Horstmann	1900 Rittenhouse

Note: There are intermittent listings for William H. Horstmann in the mid 1850s, 1879,1883, 1885 and 1893. In the latter year there are 2 listings one as a cutter and the other at house address at "the Gladstone". Unfortunately there are several William H. Horstmanns in Philadelphia during this time, one of which founded another, completely different, Horstmann Company. There is no known family connection between these Horstmanns and the Horstmanns noted in this book. However, there is Sigmund's son, William H. Horstmann (II), who played a minor role in the company in the 1870s. He died in 1879. His widow, Sarah Buy Horstmann, did not die until 1960.

NEW YORK

1828	William H. Horstmann	fringe &c manuf.	105 William
1829-34	William H. Horstmann	fringe &c	94 William
1835	William H. Horstmann	fringe &c	91 Maiden-Lane
1836	William H. Horstmann	fringe	36 Platt
1837	Horstmann & Co	military goods/fringe	36 Platt
1838-44	Horstmann & Co	military goods	8 Maiden-Lane
1845-49	Horstmann Sons & Drucker	military goods	8 Maiden-Lane
1850-51	Horstmann Bros. & Co.	military goods	8 Maiden-Lane
1851	Horstmann, William H.	fancy goods	40 Maiden Lane
1852	Horstmann, William H.	fancy goods	177 Broadway
1854	Horstmann, William H.	importer	14 Dey
1855	Horstmann, William H.	fancygoods	41 Platt
1852-68	Horstmann Bros. & Allien	military goods	8 Maiden-Lane
1869-73	Horstmann Bros. & Allien	military goods	540 Broadway
1873-76	Horstmann Bros. & Allien	military goods	7 Bond Street
1877-84	H. V. Allien & Co	military goods	7 Bond Street
1885-93	H. V. Allien & Co	military goods	738 Broadway
1894-1920	H. V. Allien & Co	military goods	734 Broadway
1920-1948	H. V. Allien & Co	military goods	227 Lexington

Note: Gustavus Drucker is listed as an importer at 8 Maiden-Lane from 1838 and at 36 Platt in 1836-37. He was thus associated with the Horstmanns before the 1845 partnership of Horstmann Bros. & Drucker. He does not appear in the directories after that firm dissolved in 1850.

Personal Addresses New York

1837-1841	William H. Horstmann	home William
1842	William H. Horstmann Jr.	home 27 Dominick
1843	William H. Horstmann Jr.	no address given
1852-53	William H. Horstmann	home 210 Livingston, Bkln.
1854	William H. Horstmann	home 207 Henry
1855	William H. Horstmann	home 209 Henry, Bkln.

Note: The relationship of this firm to the main Horstmann business is not at all clear. It would seem that William H. Horstmann Jr. was a third son and may be the "Henry Horstmann" at Mr. Beck's school. He disappears shortly after Horstmann Bros. & Allien were founded. He is apparently not the same person as William H. Horstmann II cited above in the 1870s as this was clearly a young man who had a short career with the firm and then died. Both individuals had a minimal effect on the firm's history.

Columbia Manufactory from printed label in Horstmann scrap book

51 N. 3rd Street from cards printed in May, 1848 (Horstmann label books, Historical Society of Pennsylvania)

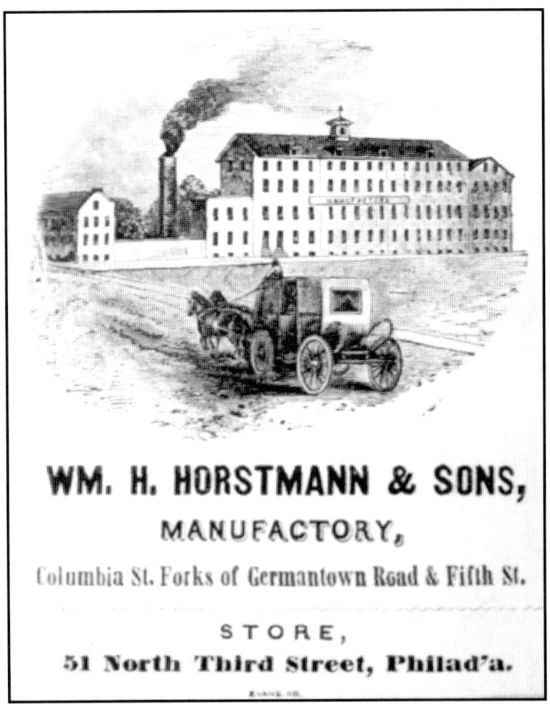

Columbia Manufactory from cards printed in May, 1848 (Horstmann label books, Historical Society of Pennsylvania)

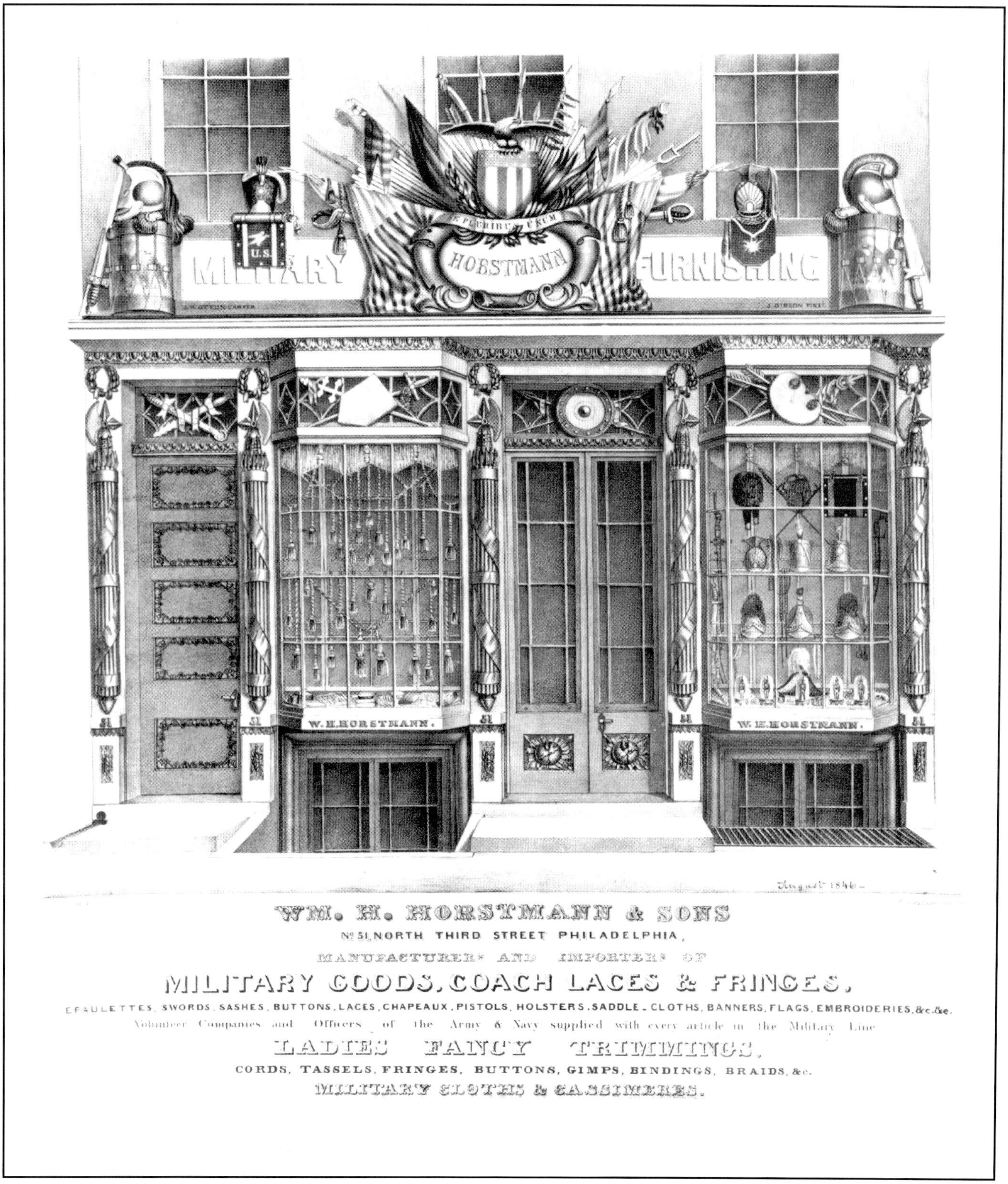

51 N. 3rd Street, August 1846 (Library Company of Philadelphia)

5th and Cherry Street factory showing original building c. 1859 and new sales rooms added during the Civil War (Horstmann scrapbooks)

MANUFACTORY.

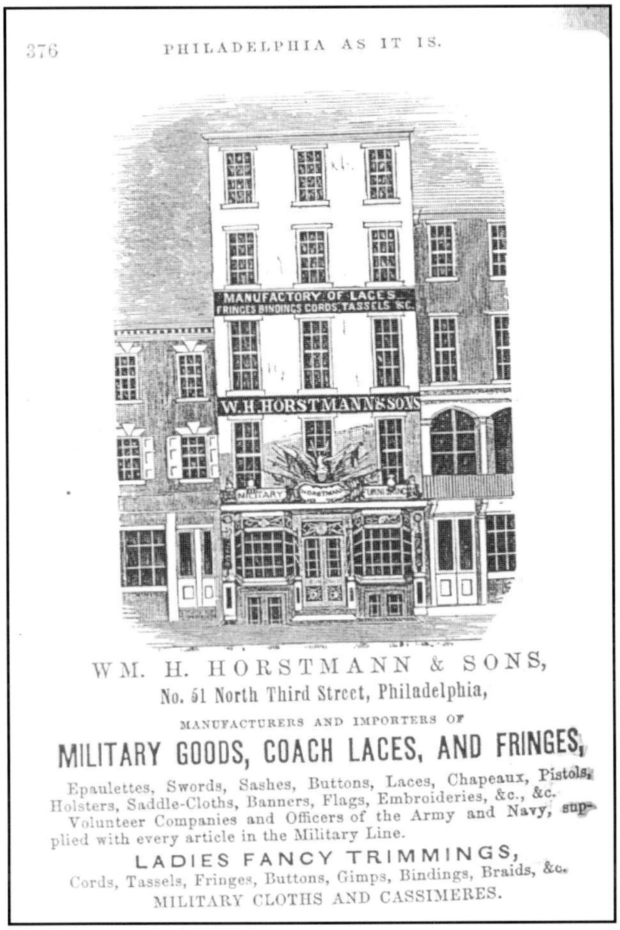

Horstmann's advertisement from A. R. Smith's *Philadelphia As It Is in 1852*

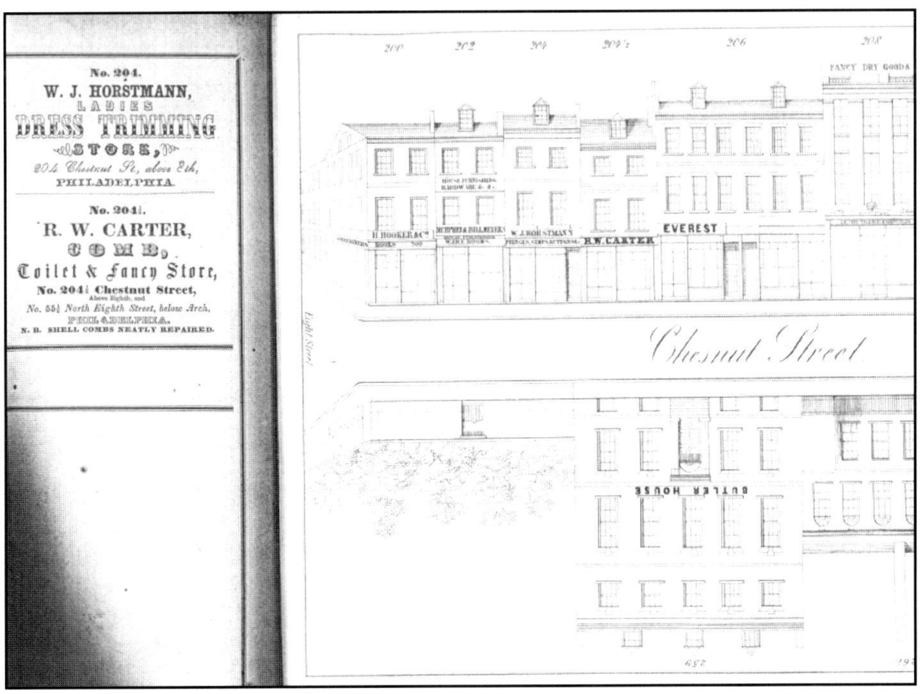

William J. Horstmann's trimming store at 204 Chestnut Street from *Rae's Philadelphia Pictorial Directory & Panoramic Advertiser*, 1851

During the second half of the 19th century it was common for insurance companies to survey the larger firms which used their policies. Horstmann was surveyed by both Hexamer and Duy & Woods:

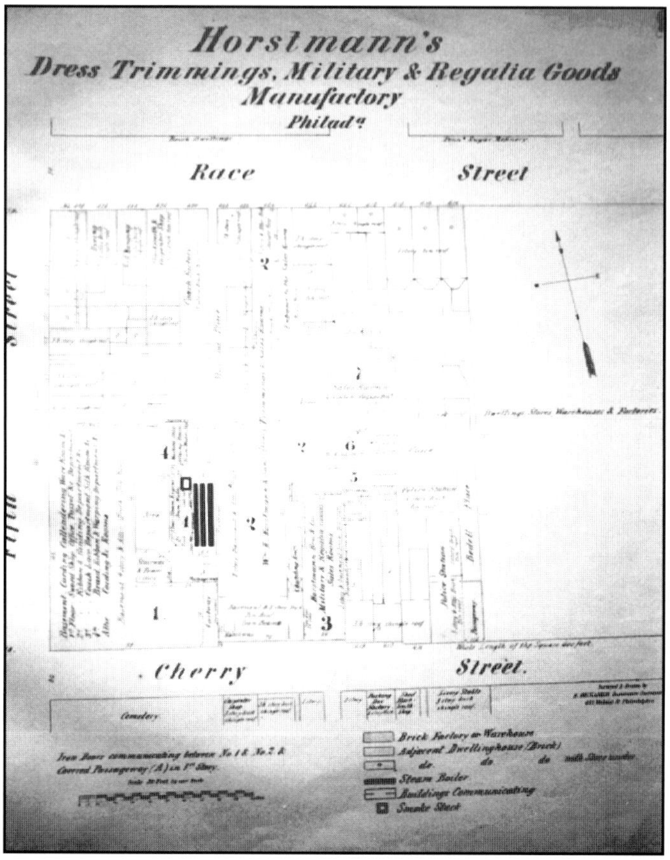

Hexamer Insurance Survey, c. 1865
(Philadelphia Free Library)

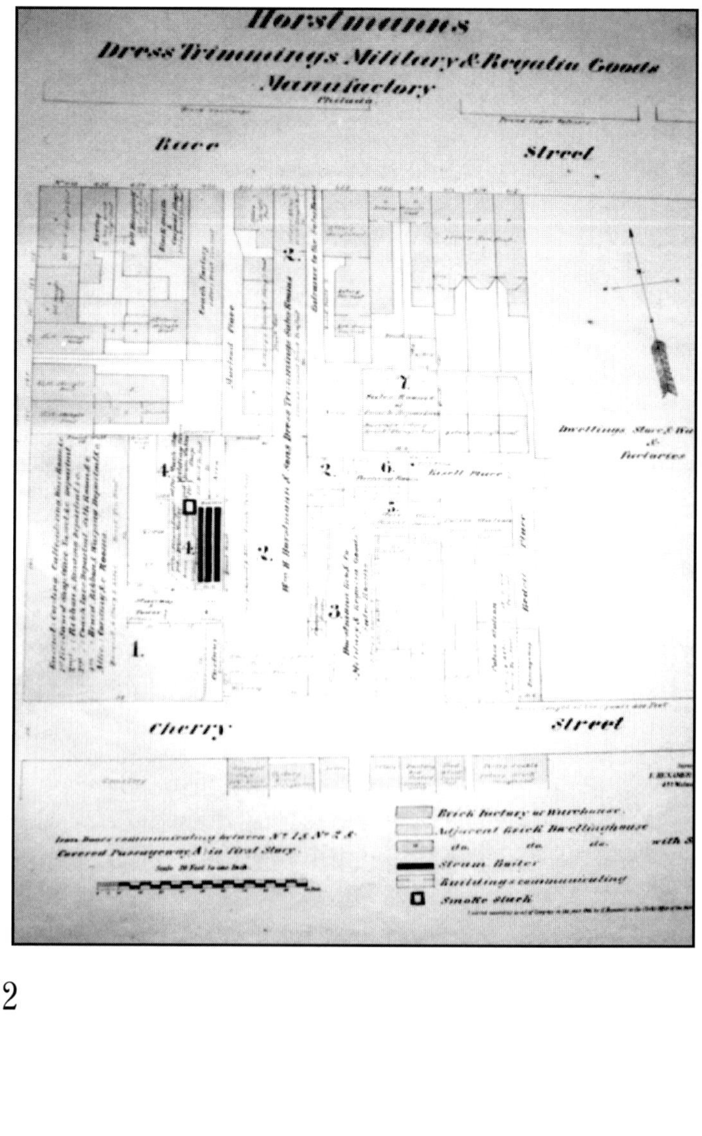

Hexamer Insurance Survey, c. 1866
(Philadelphia Free Library)

Hexamer Insurance Survey, October, 1872 (Philadelphia Free Library)

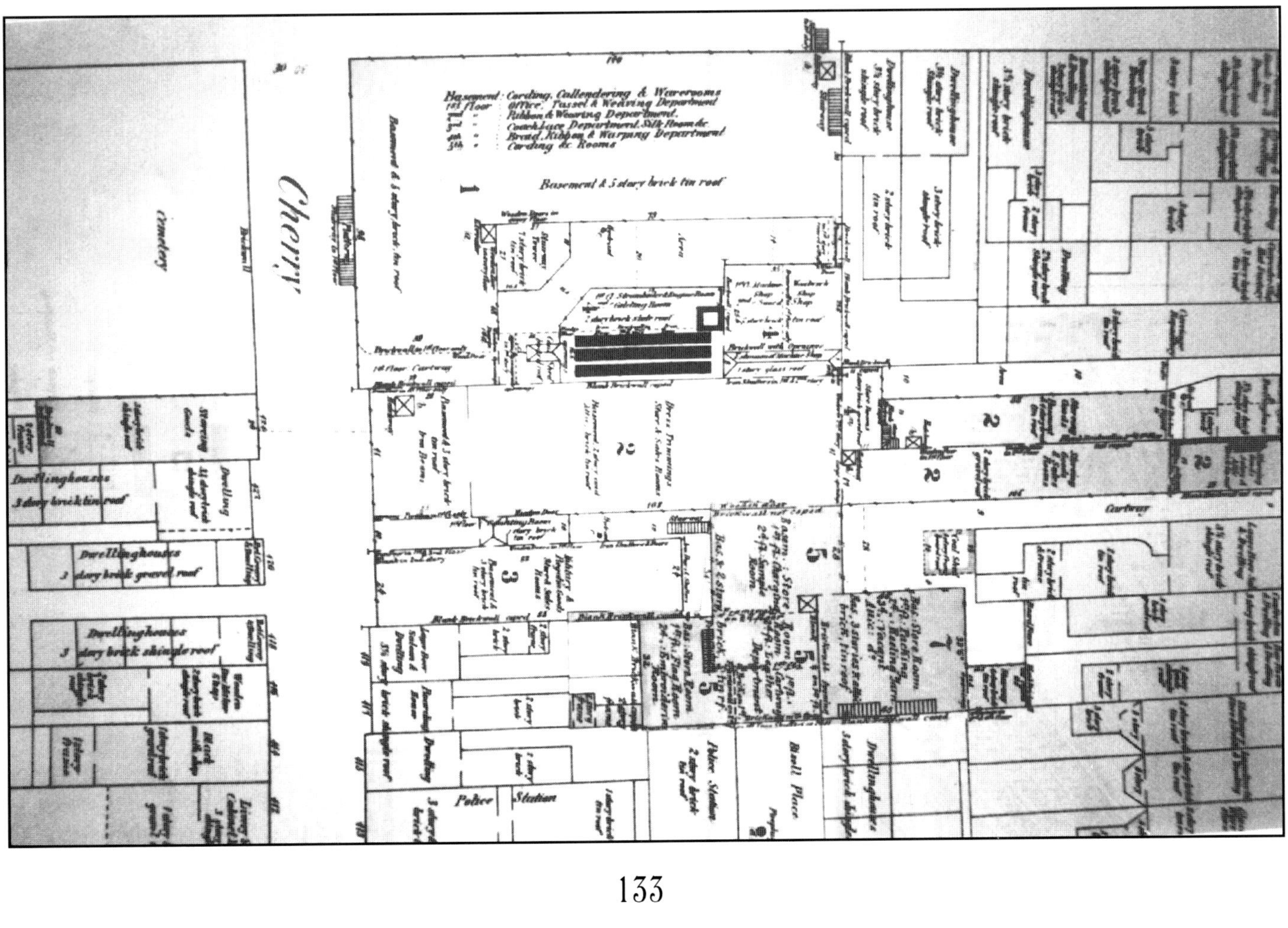

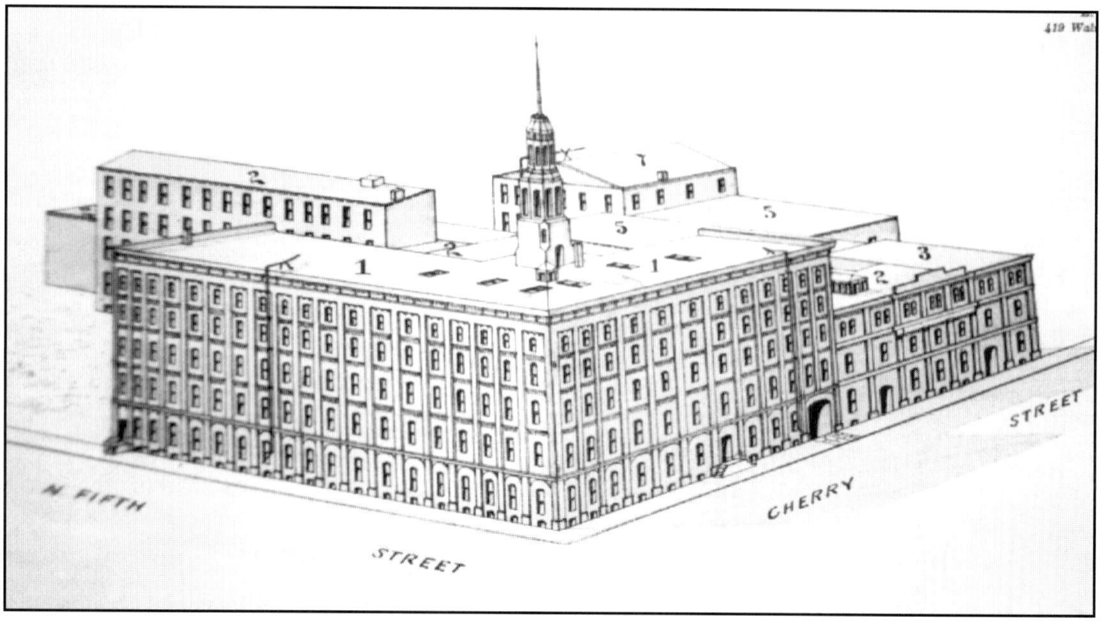

Hexamer Insurance Survey, October, 1887 (Philadelphia Free Library)

Graves of W. H. Horstmann, Sarah Horstmann, and Henrietta Horstmann located in St. Michael's Lutheran Church cemetery in Germantown. W. H. Horstmann's birth and death dates can be confirmed from the stone on the right partially turned over by vandals.

Gravemarker: William J. and Emma Louisa Horstmann, Laurel Hill Cemetery, Philadelphia

Cemetery marker for Sigmund H., Elizabeth, and Sallie Horstmann in Laurel Hill Cemetery, Philadelphia

Gravemarker: F. Oden Horstmann, Laurel Hill Cemetery, Philadelphia

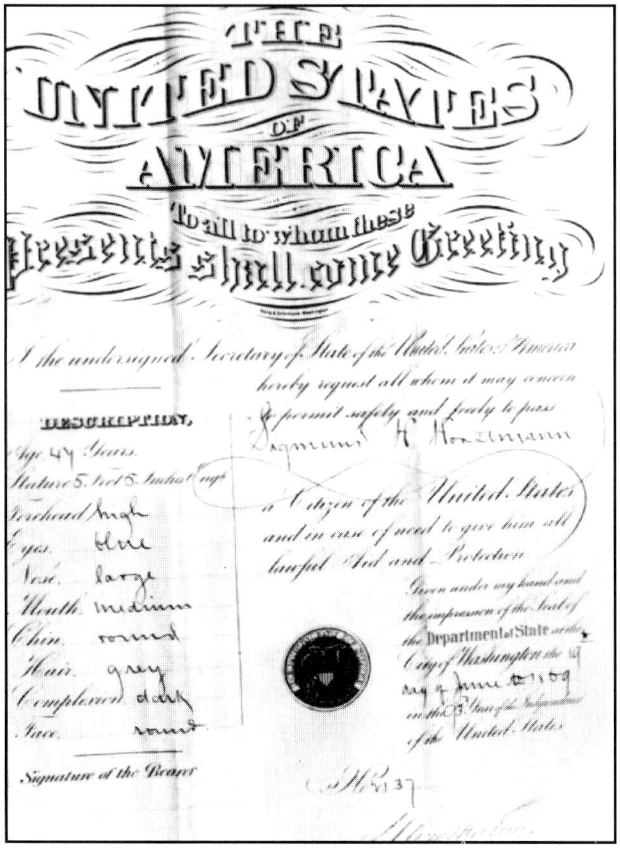
THE
UNITED STATES
OF
AMERICA
To all to whom these Presents shall come Greeting

I the undersigned Secretary of State of the United States of America hereby request all whom it may concern to permit safely and freely to pass Sigmund H. Horstmann a Citizen of the United States and in case of need to give him all lawful Aid and Protection

DESCRIPTION.
Age 47 Years
Stature 5 Feet 5 Inches Eng
Forehead high
Eyes blue
Nose large
Mouth medium
Chin round
Hair grey
Complexion dark
Face round

Signature of the Bearer

Given under my hand and the impression of the Seal of the Department of State at the City of Washington the 4 day of June 1870 in the 94 Year of the Independence of the United States

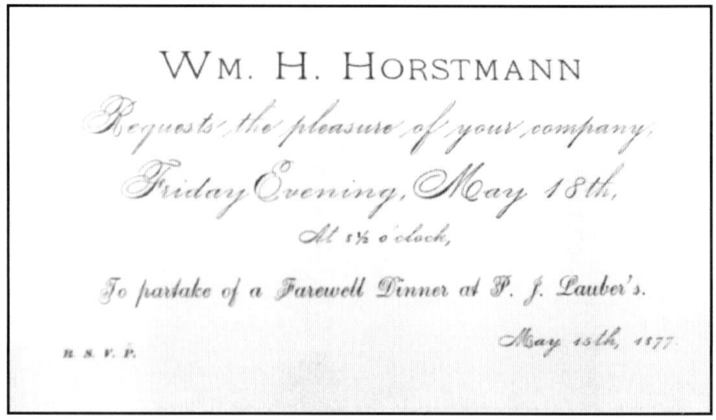
WM. H. HORSTMANN
Requests the pleasure of your company,
Friday Evening, May 18th,
At 1½ o'clock,
To partake of a Farewell Dinner at P. J. Lauber's.

R. S. V. P.

May 15th, 1877

Invitation to Farewell Dinner for William H. Horstmann (II) in 1877 (Horstmann Papers, Historical Society of Pennsylvania)

Passport for Sigmund Horstmann to visit Italy in 1870 (Horstmann Papers. Historical Society of Pennsylvania)

Gravemarker: Walter Horstmann, Laurel Hill Cemetery, Philadelphia

Gravemarker: F. Oden Horstmann, Jr., Laurel Hill Cemetery, Philadelphia

Stevensgraph made by the Horstmann Company to honor the memory of Sigmund Horstmann. This technique manufactured a sewn portrait or scene in woven silk and was considered a triumph of the textile trade. (Author's collection)

Walter Horstmann (1857-1922). The last Horstmann descendant involved in managing the business.

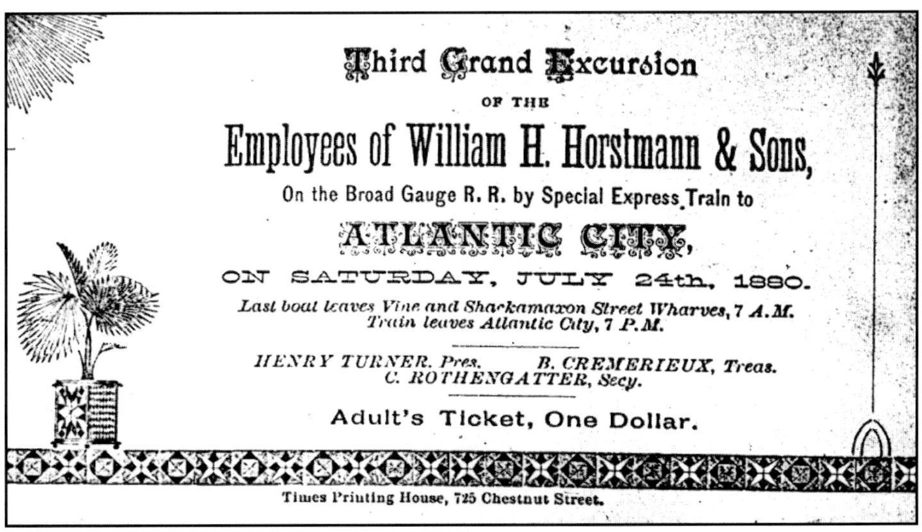

Third Grand Excursion

OF THE

Employees of William H. Horstmann & Sons,

On the Broad Gauge R. R. by Special Express Train to

ATLANTIC CITY,

ON SATURDAY, JULY 24th, 1880.

Last boat leaves Vine and Shackamaxon Street Wharves, 7 A.M.
Train leaves Atlantic City, 7 P.M.

HENRY TURNER, Pres. B. CREMERIEUX, Treas.
C. ROTHENGATTER, Secy.

Adult's Ticket, One Dollar.

Times Printing House, 725 Chestnut Street.

Ticket for the Horstmann Employees Annual Excursion to Atlantic City in 1880. This excursion was conducted for a number of years in the late 19th century. (Horstmann scrapbook)

Established 1815. Factory and Salesrooms, Fifth and Cherry Streets, Philadelphia.

Philadelphia, January 1st, 1877.

The Co-partnership heretofore existing between the firms of HORSTMANN BROS. & CO., Philadelphia, and HORSTMANN BROS. & ALLIEN, New York, was dissolved by mutual agreement on Dec. 31st, 1876.

F. ODEN HORSTMANN will attend to the liquidation of the affairs of Horstmann Bros. & Co., Philadelphia, and HENRY V. ALLIEN to those of Horstmann Bros. & Allien, New York.

Very Respectfully,

F. O. HORSTMANN,
HENRY V. ALLIEN.

1877 notice proclaims the dissolution of Horstmann Brothers & Allien (Horstmann scrapbook)

The old Horstmann Bros. & Allien firm was replaced by the Henry V. Allien Company (Horstmann scrap book)

Philadelphia, January 1, 1877.

DEAR SIRS:

In calling your attention to the notice of dissolution on the opposite page, we desire to add that our business of

MANUFACTURING AND IMPORTING

MILITARY, SOCIETY REGALIA

AND

THEATRICAL GOODS,

Banners, Flags, &c.

will be continued in Philadelphia as heretofore.

Our large facilities for manufacturing and importing goods will enable us to offer the best inducements to the trade.

Soliciting a continuance of your past favors, we remain,

Yours, Respectfully,

Horstmann Brothers & Co.

FIFTH AND CHERRY STREETS,
PHILADELPHIA.

N. B.—MAIL ORDERS AND INQUIRIES WILL RECEIVE PROMPT ATTENTION.

No 4/3

American Academy of Music.

Stockholders' Ticket

LADY'S TICKET.

For the Season ending July 31st 1875

Issued on Certificate No 3071

To Bessie T. Horstmann

FORFEITED IF USED BY ANY OTHER PERSON.

To be shown at the door.

Secretary

Ticket for Bessie Horstmann (Elizabeth Trotter Horstmann) for the American Academy of Music, 1975 (Courtesy: Jackie Eubanks)

GREAT CENTRAL FAIR FOR THE U. S. SANITARY COMMISSION.

DEPARTMENT OF MILITARY GOODS.

As the time approaches in which we must make the necessary arrangements for the display and sale of the articles contributed, and also, the better to enable us to form an idea as to the space required for our department in the Fair Buildings, we take this method of respectfully referring to the Circular addressed to you a short time since by our Committee, and would ask of you a reply at your earliest convenience, as to your intention to contribute to the GREAT CAUSE OF HUMANITY, by articles of your manufacture, or in such other form as may be more agreeable to you. Donations in money are also solicited, and receipts will be returned by the Treasurer to the credit of our department.

Goods should be addressed:

A. R. McHENRY,
GREAT CENTRAL FAIR, PHILA.
For Department of Military Goods from A. B. C.

Send *invoice* and *bill of lading* or *receipt* of *Railroad or Express Co.*, directed as above, or to the Chairman of this Committee.

Mark each article at a reasonable value.

On receipt of each article a proper acknowledgment will be sent duly attested.

WM. J. HORSTMANN, *Chairman*,
N. E. cor. 5th and Cherry St.

Broadside showing William J. Horstmann as Chairman of the Department of Military Goods in the Great Central Fair held in Philadephia in 1864. W. J. Horstmann would go on to serve in the U. S. Centennial Commission. (Horstmann scrapbook)

New-York, April, 1873.

Dear Sirs:

We beg to inform you that we have
REMOVED
from 540 Broadway, to the new and commodious store
No. 7 Bond St.,
a few doors from Broadway, opposite the Grand Central Hotel.

Very Respectfully,

HORSTMANN BROS. & ALLIEN,
(Formerly 8 Maiden Lane.)

38 Rue Meslay, Paris.

MANUFACTURERS AND IMPORTERS OF
MILITARY, THEATRICAL, CHURCH & REGALIA GOODS.
SILK, MUSLIN AND BUNTING FLAGS,
BANNER SILKS AND BUNTING.

1873 broadside showing Horstmann Bros. & Allien moving to 7 Bond Street, New York (Horstmann scrapbook)

Share No. 11 *Philadelphia, Nov. 23rd 1870*

This is to Certify, *That Miss Mamie Horstmann*
has paid $ 1 for one Share in the
Schoolboys' Circulating Library,
thereby being entitled to draw books from the Library according to the Rules and By-Laws.

W. Horstmann Librarian. *Wm. H. Horstmann* President.

Schoolboys' Circulating Library share signed by William H. Horstmann (II) in 1870 (Author's collection)

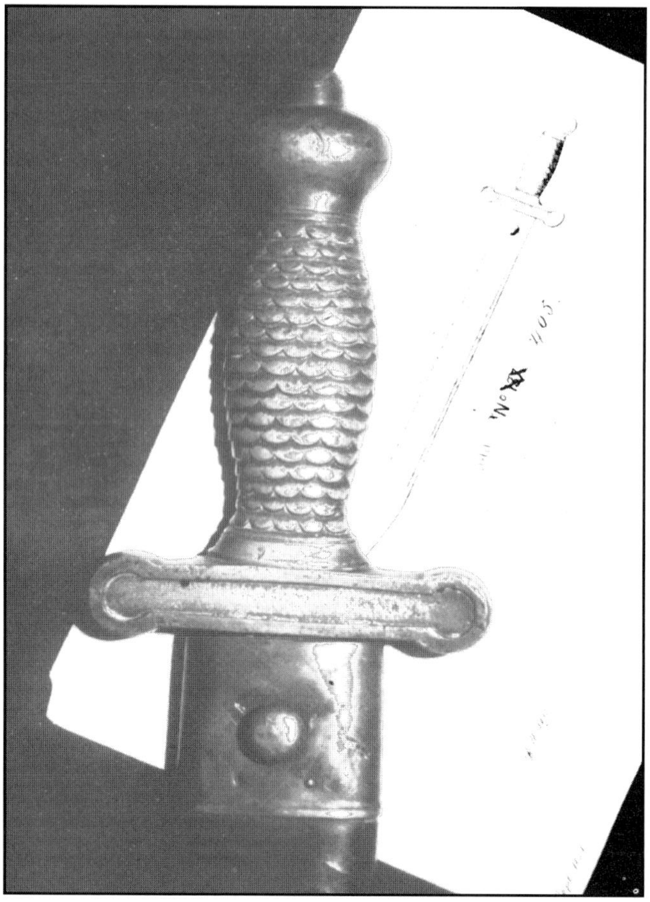

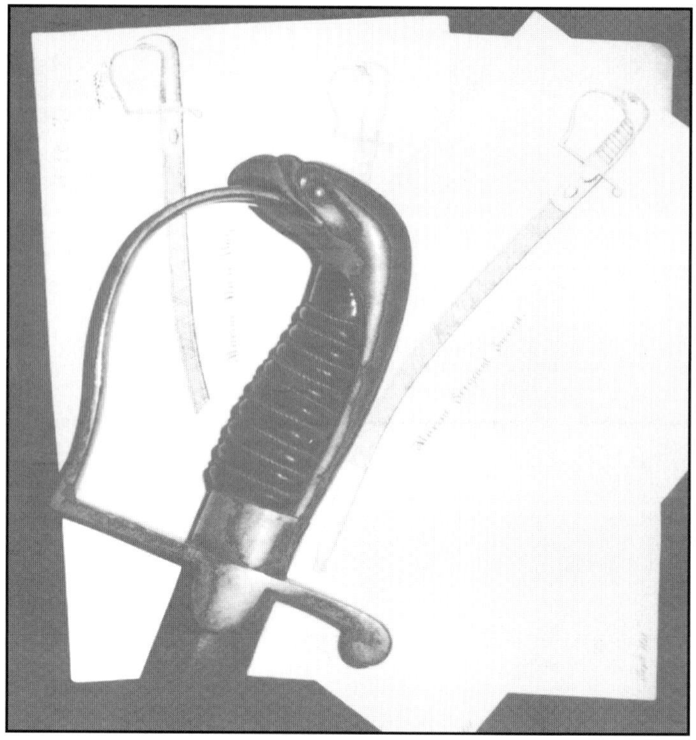

Work drawings for Widmann/Horstmann swords. After F. W, Widmann's death in April, 1848, his business was sold to Horstmann. Accompanying it were at least two of Widmann's workers, Louis Loemmel and Julius Knocke, who were employed by Horstmann from May, 1848. One of them probably kept these work drawings which span at least 1846-1851. At left, Artillery NCO sword with photo of actual Horstmann drawing. At right, hilted Marine Sergeant's sword with its work drawing. (Workbook and swords, courtesy Norm Flayderman, pages 140-143)

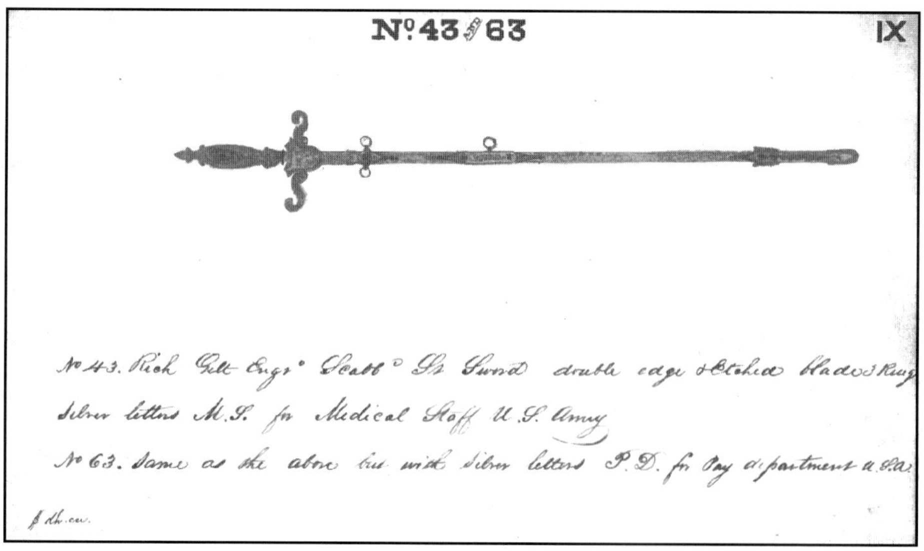

Medical Service sword drawing

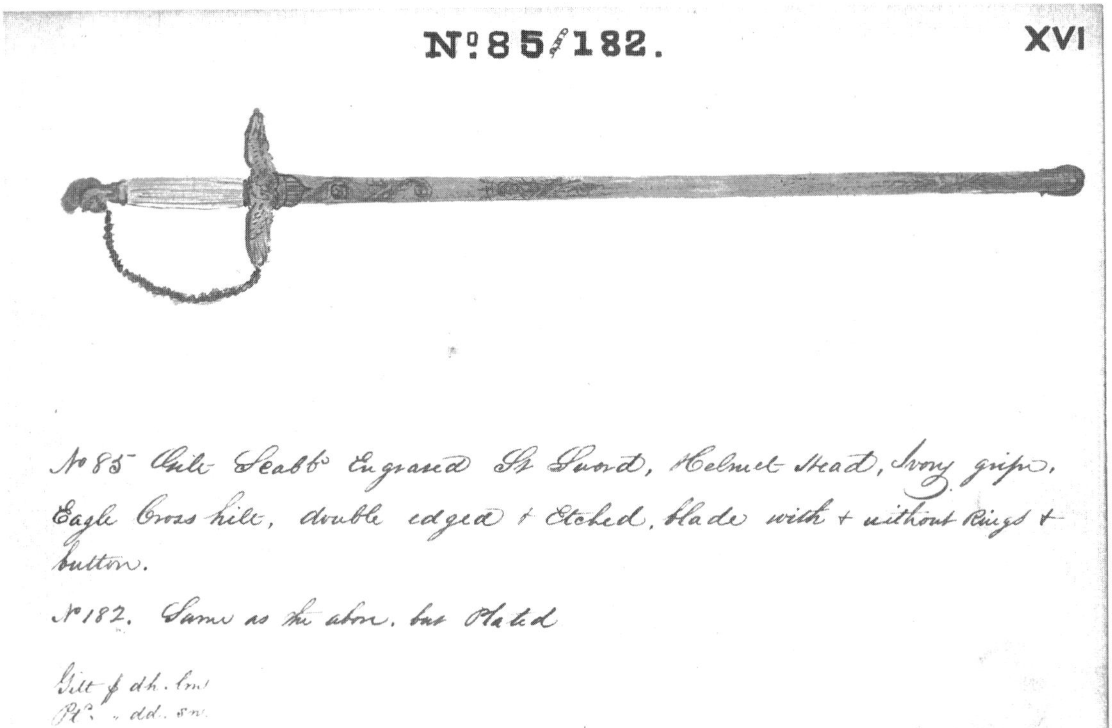

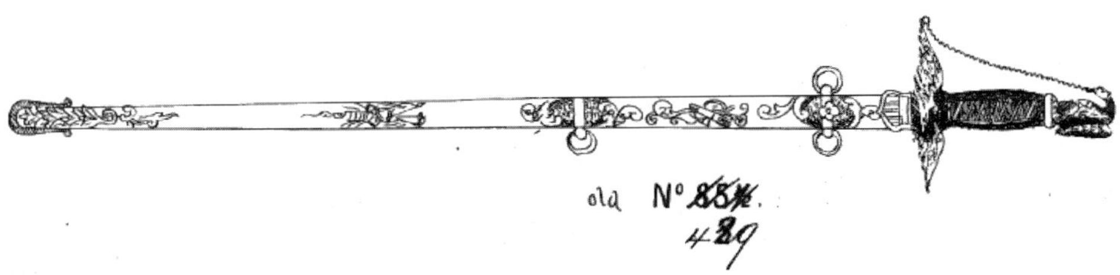

Staff Officers' swords, one of which has been renumbered 489, probably to conform to Horstmann's products numbering which gave his sword patterns numbers between 400 and 550.

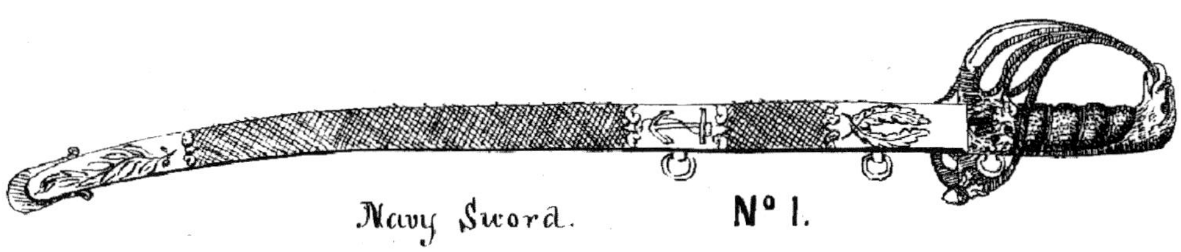

Navy sword

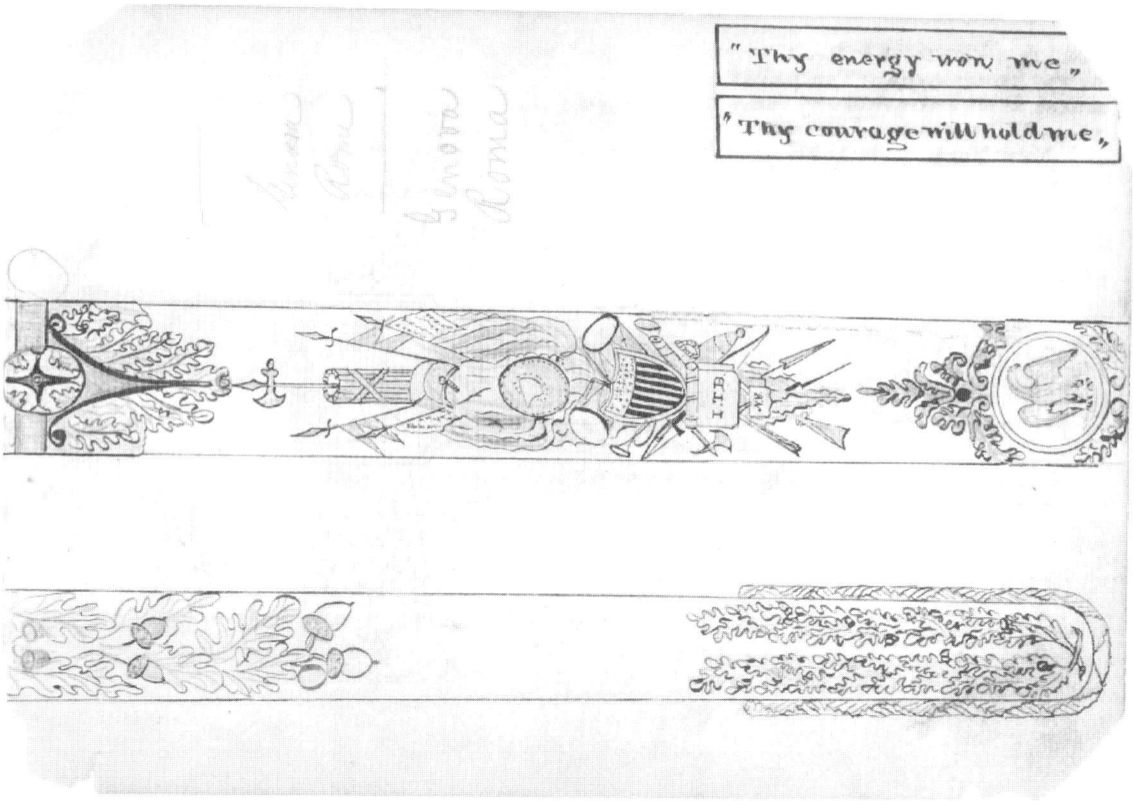

Details of drawings for a sword given to Colonel Elijah Ward on July 4, 1848. This was three months after Widmann's death.

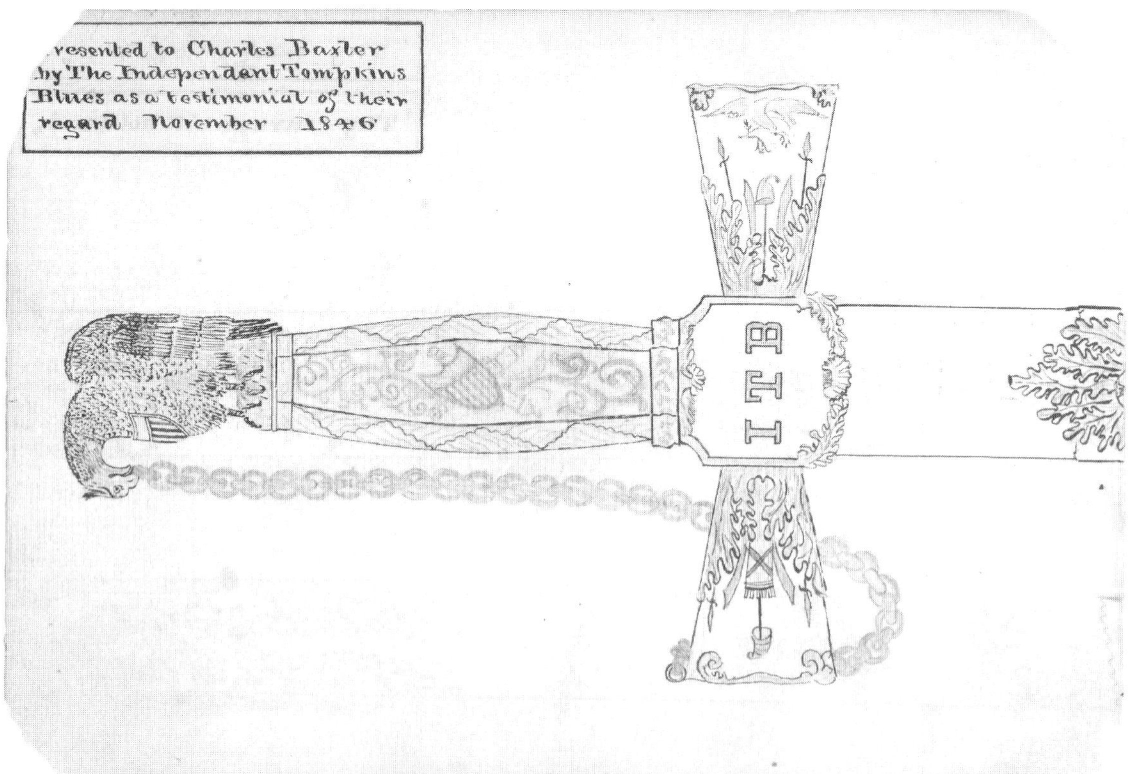

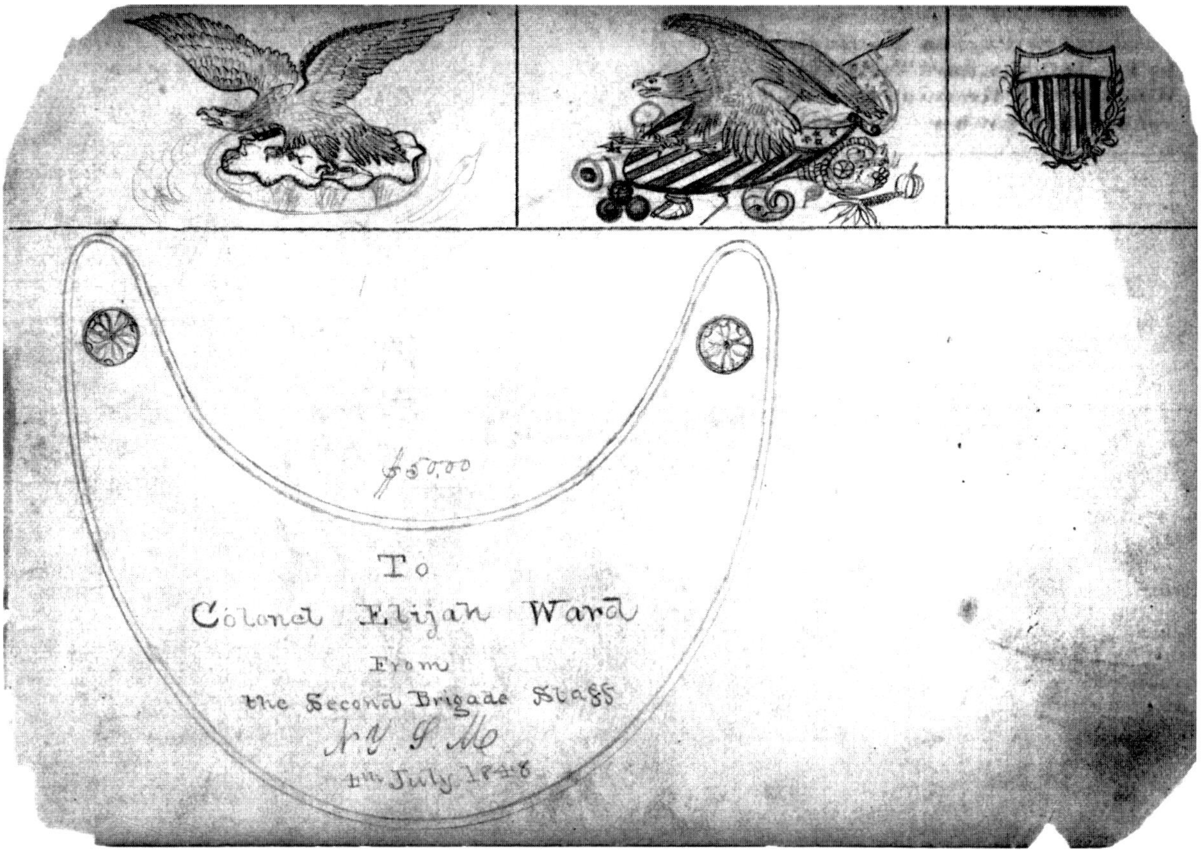

Details of drawings for a sword given to Charles Baxter from the Tompkins Blues of New York, 1846

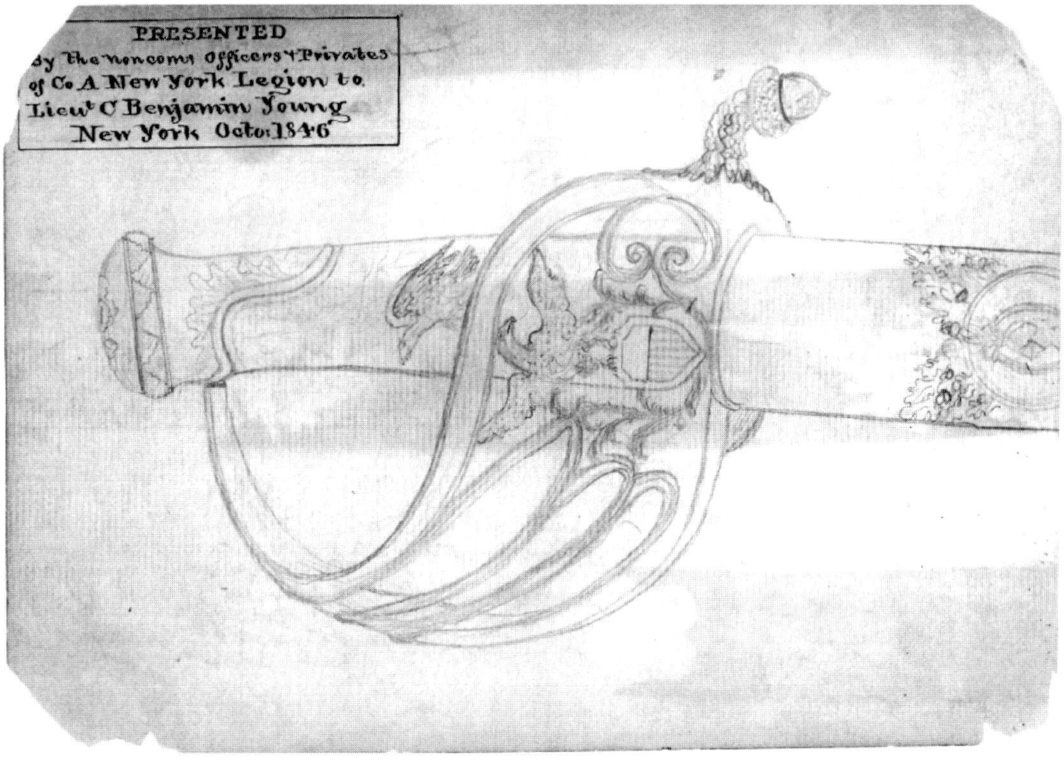

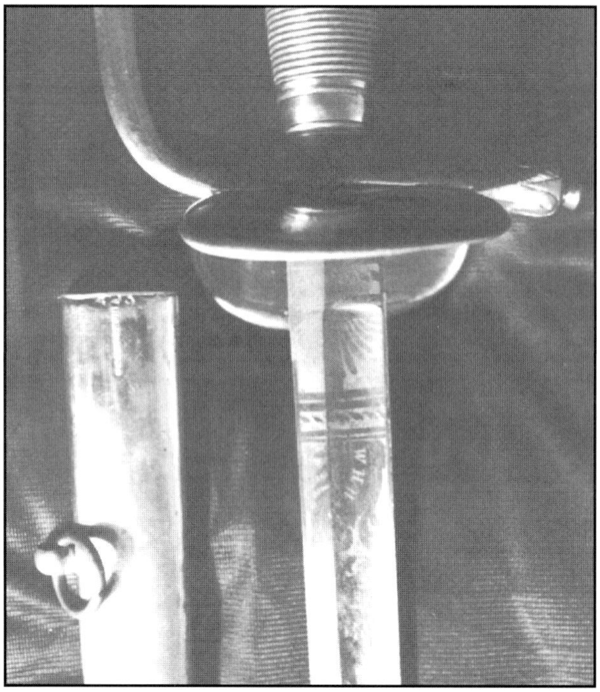

Model 1840 foot officer's sword mounted by Horstmann c. 1850 for militia use. The blade is marked W. H. H. Co./ Philadelphia and the scabbord "Horstmann 425". The sword is almost identical to one carried by General Paul Applebach c. 1854, now in the Mercer Museum private collection.

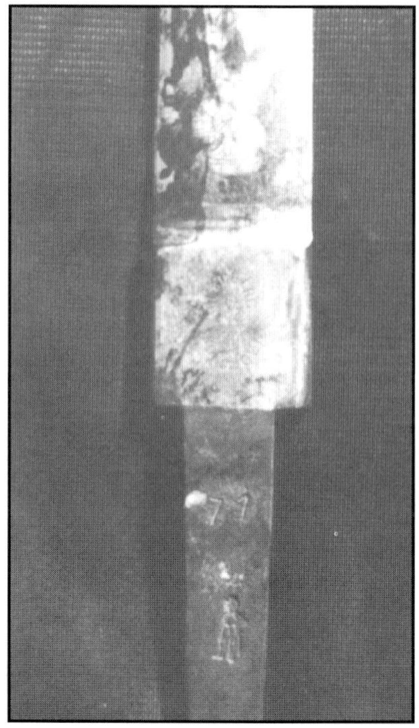

Detail of German blade believed made for Horstmann to mount in militia staff officers' swords (Author's collection)

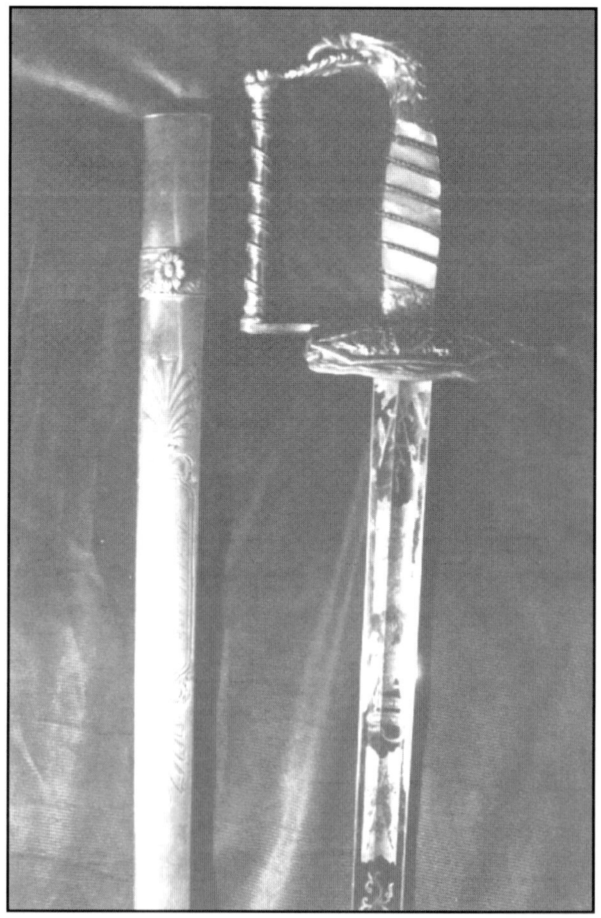

Sword presented to Peter Fritz by the National Greys, July 4, 1836. Scabbard is marked F.W. Widmann/ Philada. Portrait of Fritz holding the sword is on page 145.

Cadet sword marked "Horstmann/ & Sons/ Philadelphia" c. 1860 (Private collection)

Horstmann Products

Horstmann's military customers were the U. S, Army and the volunteer militia. Shown is a portrait of Peter Fritz, Captain of the National Grays of Philadelphia holding the Widmann made presentation sword given to him by the company on July 4, 1836. This sword is pictured on page 144. (Portrait, Historical Society of Pennsylvania. Sword: Private collection)

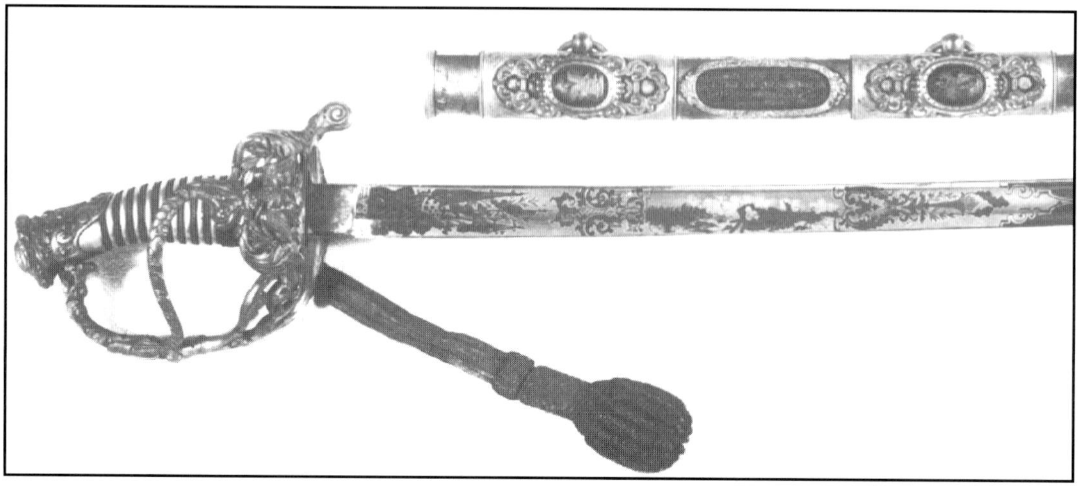

Sword presented to Major J. K. Bolton of the 54th Pennsylvania Volunteer Infantry in 1865. The sword features a high quality German made gold washed blade marked "William H. Horstmann & Son" along with a Clauberg/Solingen manufacturer's mark. Horstmann's blades were almost always German imports. (Courtesy: State Museum of Pennsylvania)

Sword presented to Brevet Major General Galusha Pennypacker of Westchester, PA, c. 1865. Blade is marked "William H., Horstmann Sons", nominally out of the military business in 1859 but which in fact conducted an extensive business during the Civil War (Courtesy: Chester County Historical Society)

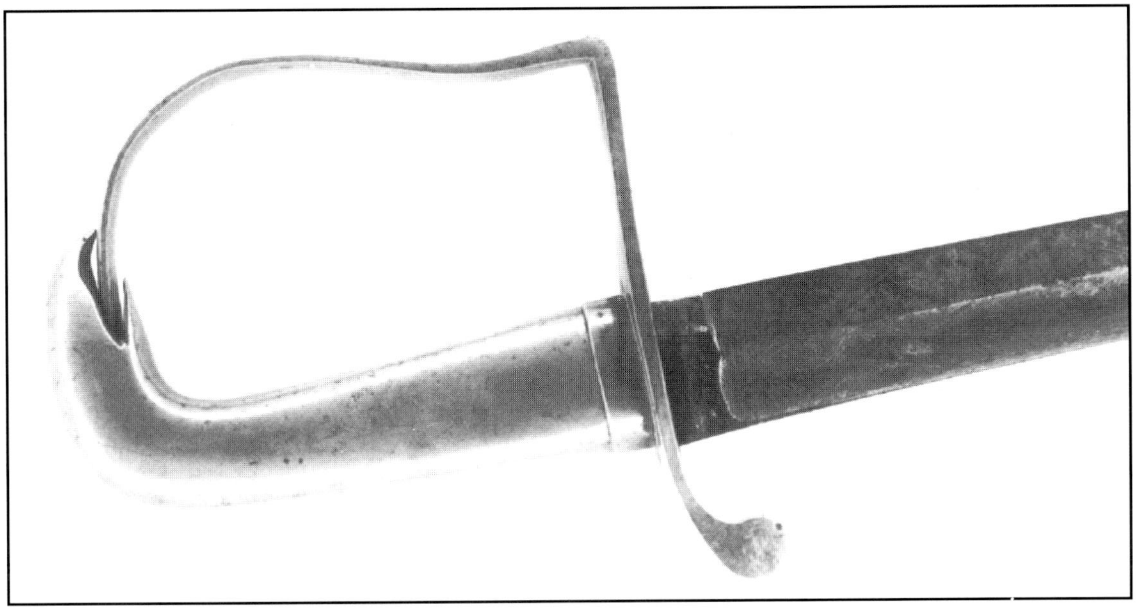

U. S. Marine Musician's sword, c. 1857. The sword was assembled by Horstmann but still shows many of the earlier Widmann design features such as the slender stylized eagle head. (Courtesy: State Museum of Pennsylvania)

Model 1860 Staff Officer's sword presented to Colonel Theodore Wiedershein, 1st Regiment, National Guard of Pennsylvania in 1877. The William H. Horstmann & Sons mark appears on this military piece long after that firm supposedly devoted itself only to civilian products. (Courtesy: First Regiment Museum, Philadelphia)

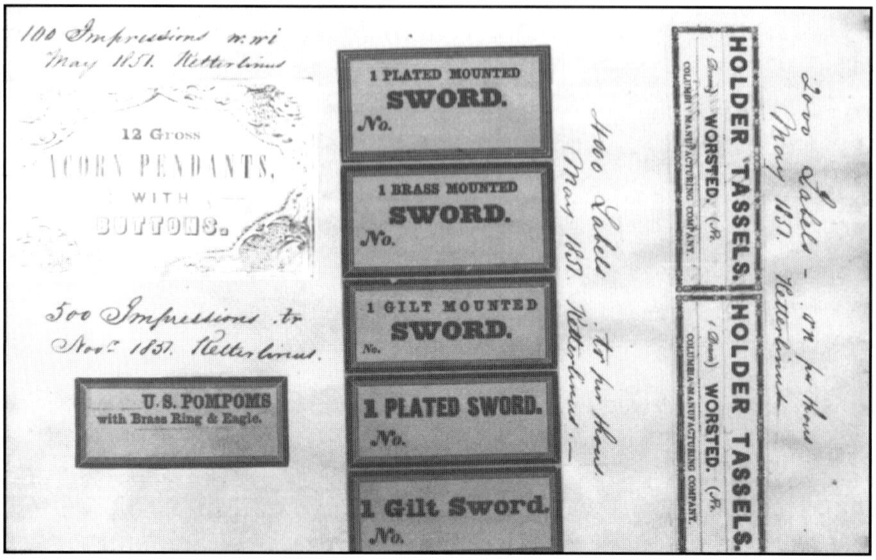

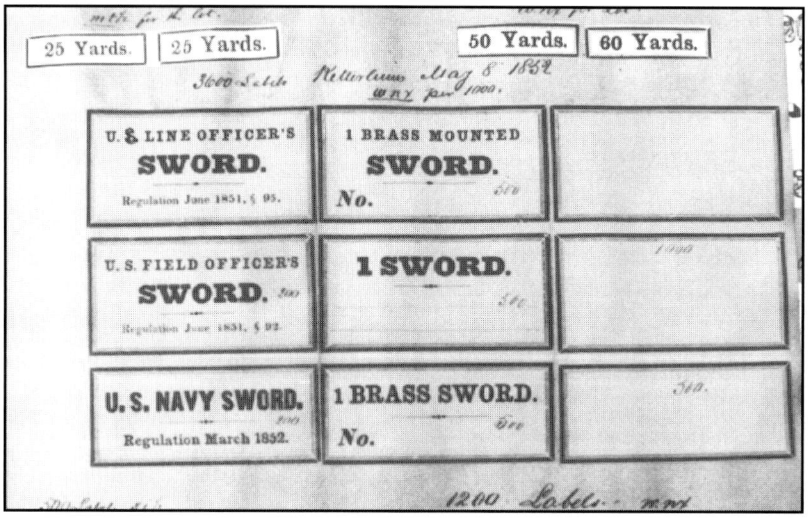

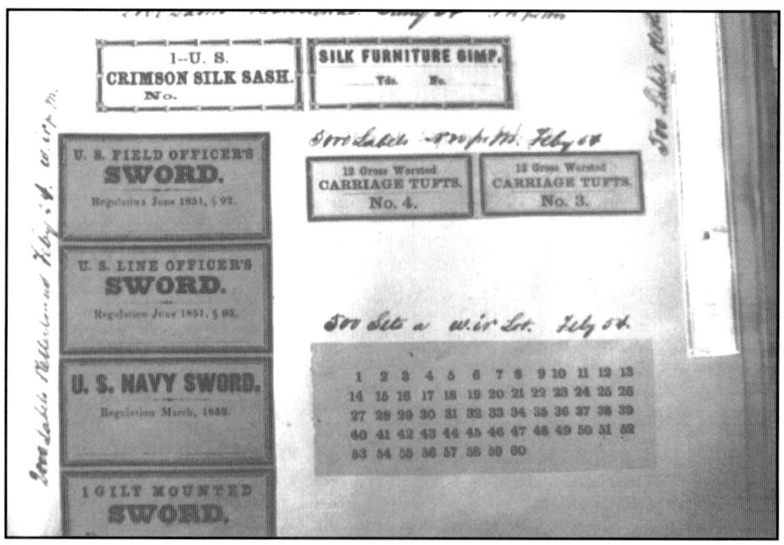

Labels for Horstmann's swords ordered May, 1851; May 8, 1852; and February, 1854. (Horstmann sample book, Historical Society of Pennsylvania)

Pages 44 through 48 of Horstmann's 1877 catalog showing Horstmann's product line of swords. Note the numbers associated with each sword with earlier numbers showing designs inherited from Widmann.

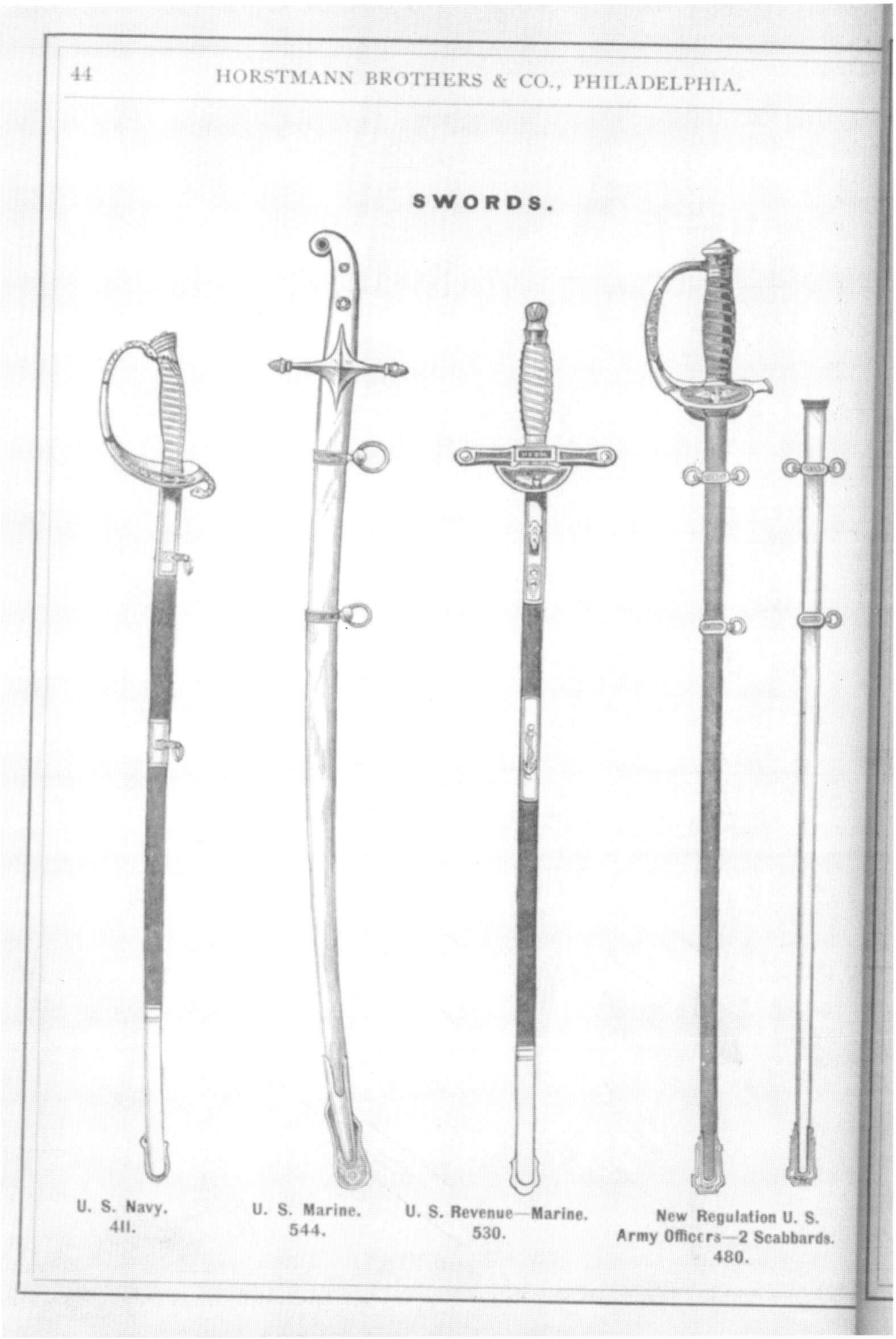

SWORDS.

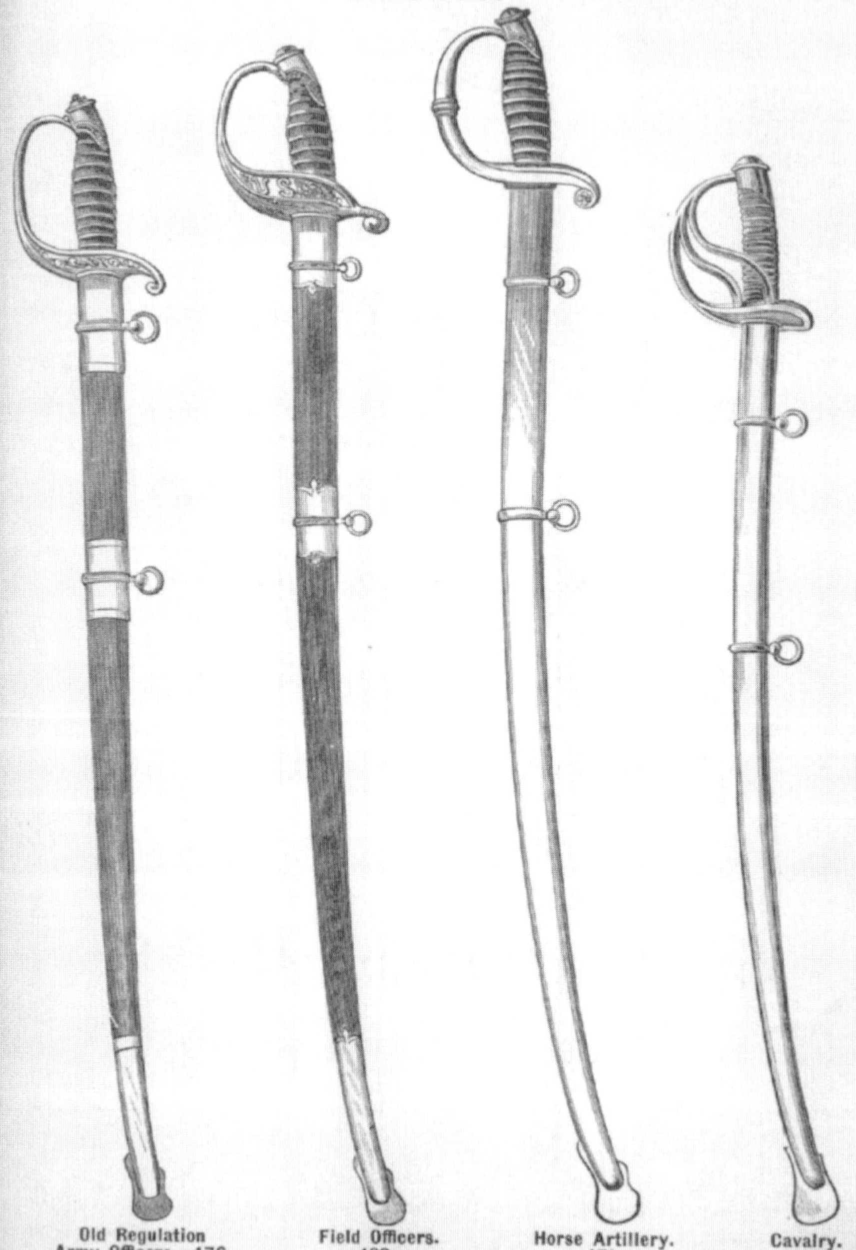

Old Regulation Army Officers—476. | Field Officers. 432. | Horse Artillery. 471. | Cavalry. 433.

SWORDS.

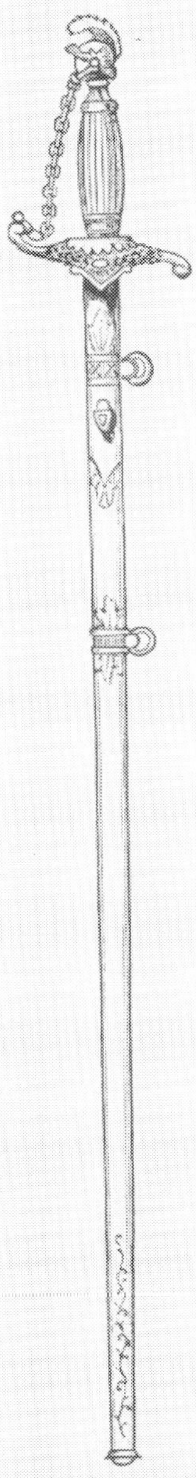

Gilt Scabbard,
Band—427.

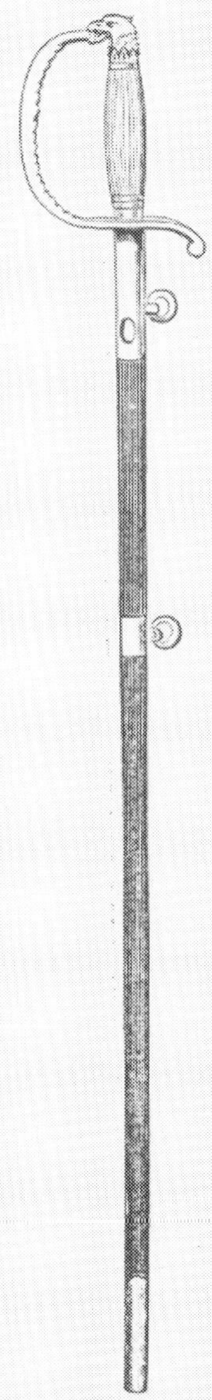

Brass Mounted,
Cadet—401.

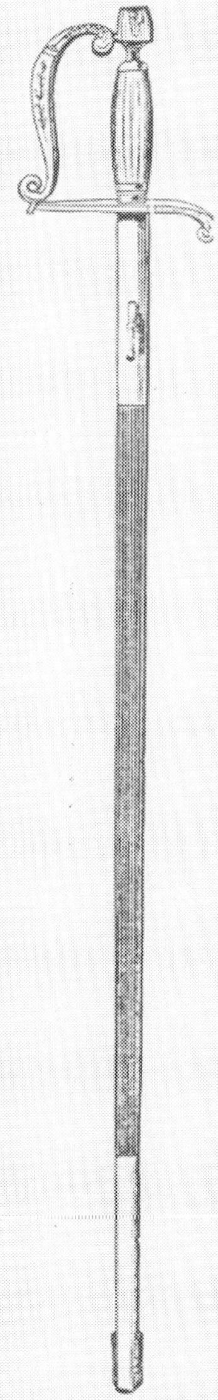

Brass Mounted,
Cadet—402.

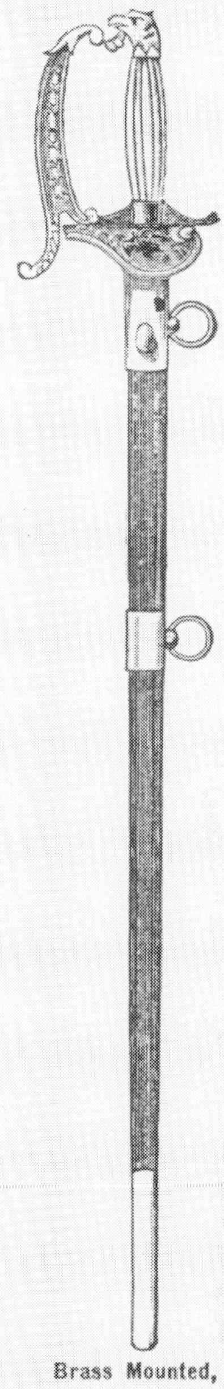

Brass Mounted,
Cadet—402 B.

SWORDS.

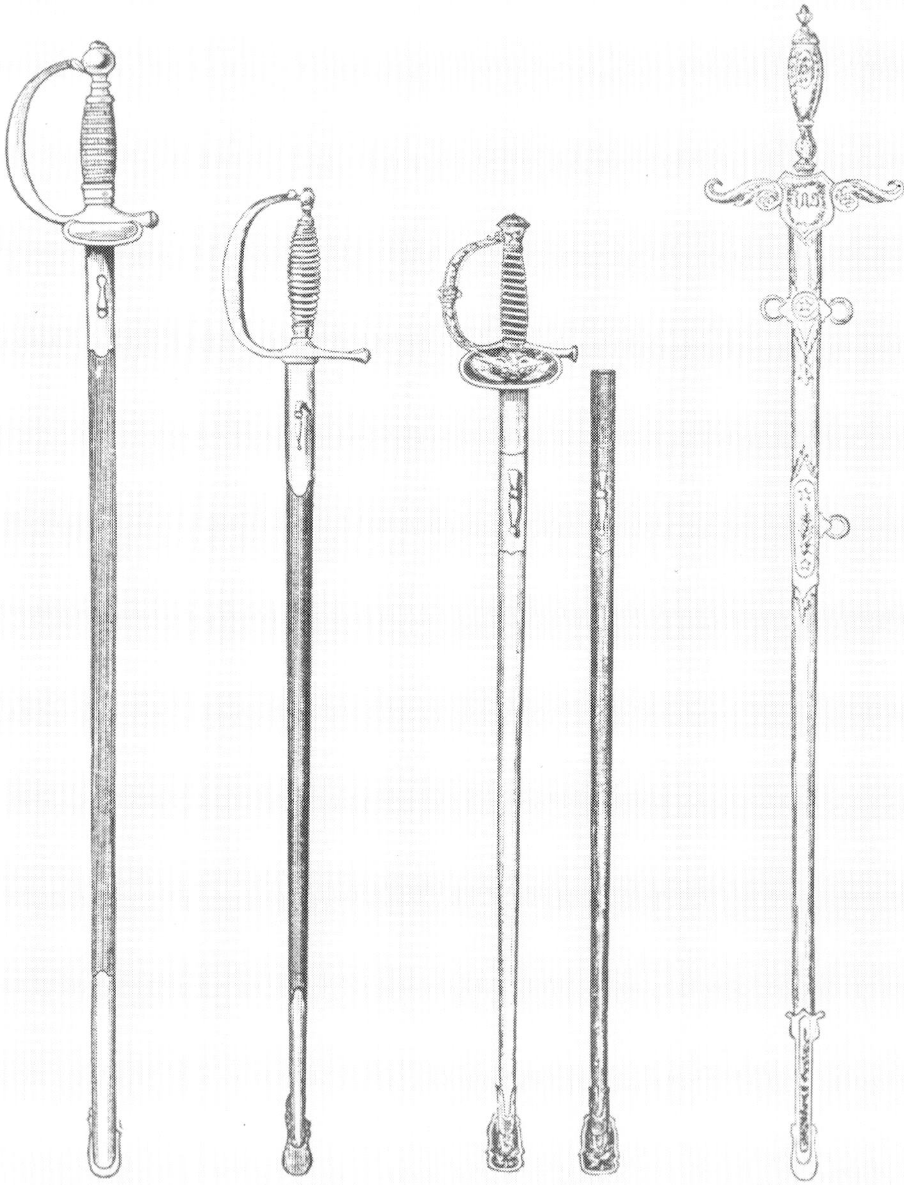

U. S. Sergeant's. 1000.

U. S. Music. 495.

U. S. Non-Commissioned, 2 Scabbards—547.

Gilt Scabbard, Pay and Medical—440.

SWORDS.

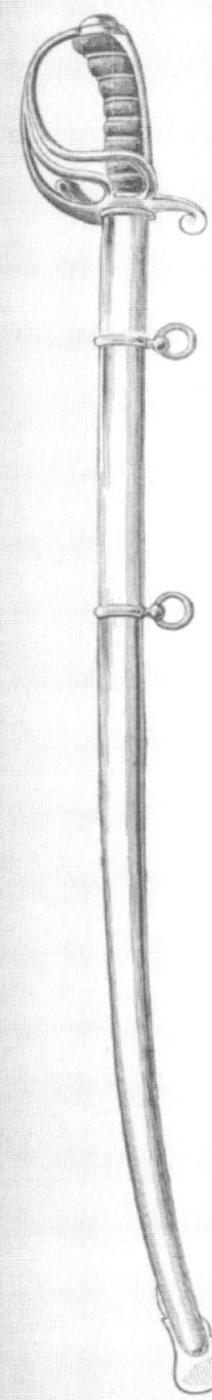

First Troop, Philad'a City Cavalry—438.

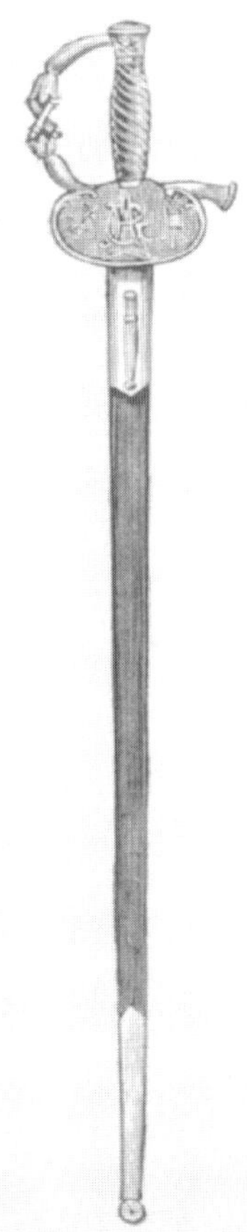

Gilt Mounted—G. A. R.

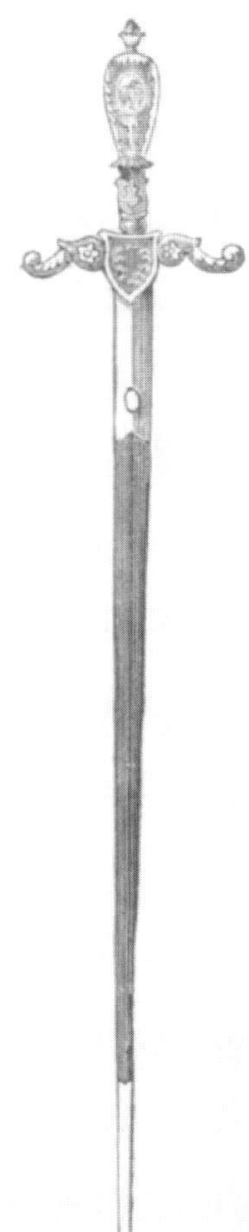

Veteran Guard, Rapier—549.

Gilt Scabbard, Band—423.

A few of the many fraternal swords assembled or sold by Horstmann

Eagle buckle marked "Horstmann" c. 1845-1900

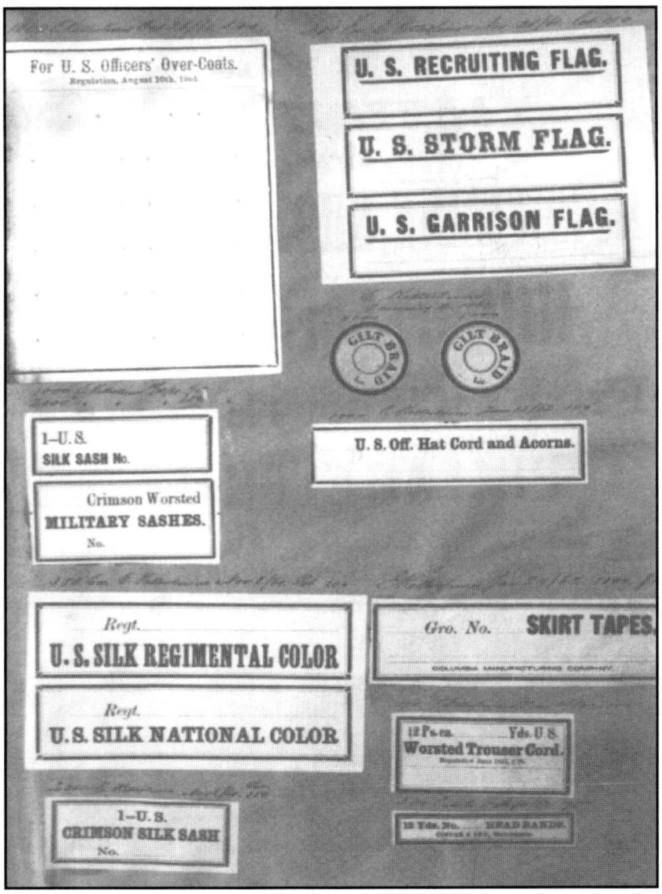

1861 labels for regimental colors and officers' overcoat buttons (Horstmann scrapbooks)

Page 58 from Horstmann's 1877 catalog showing buckles including the buckle shown above

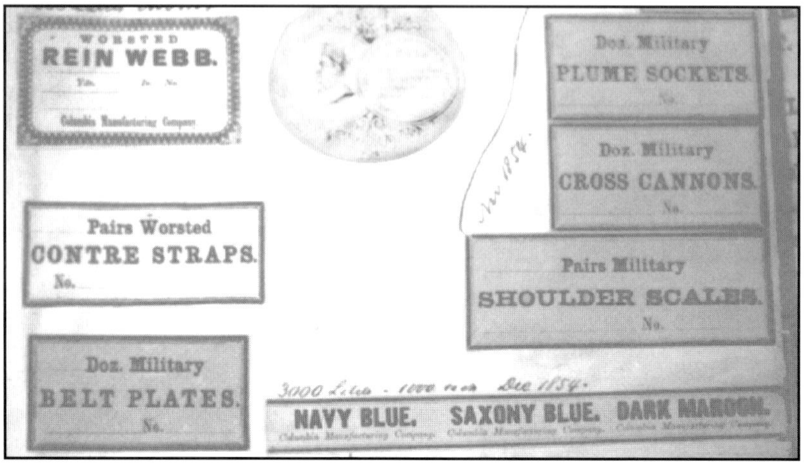

Labels printed by Horstmann for swords, eagle insignia, and such civilian products as shade tassels, fringe, and tufts, September, 1854-May 1856 (Horstmann scrapbook)

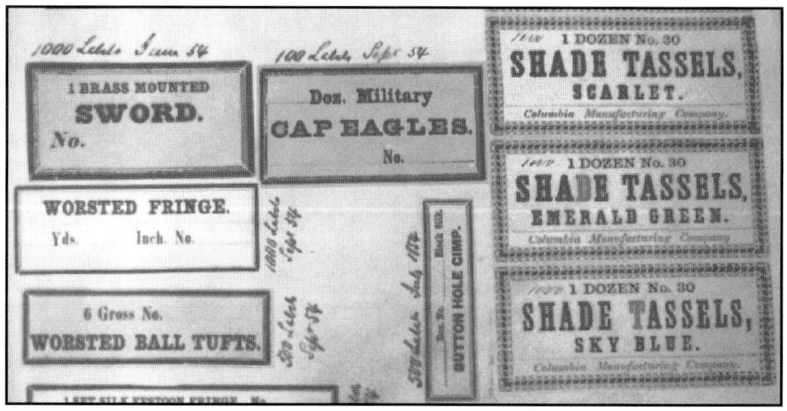

Label for chapeau de bras, Horstmann Bros. & Allien c. 1865

Wm. H. HORSTMANN & SONS

MANUFACTORY & SALES ROOMS

Cor. of Fifth & Cherry Streets

PHILADELPHIA.

Dear Sir:

We have now in store, our Spring Stock of

LADIES' DRESS & CLOAK TRIMMINGS

and call your particular attention to the following articles:

Wide Silk and Union Beltings, 2—2½ inches,
Plain and Fancy Buckles,
Square Buttons,—Gilt, Jet and Silk,
Ball Buttons,—Gilt, Jet, Silver and Alabaster,
Ornaments and Bugle Gimps,
Silk Dress Cords and Tassels.

OUR LINE OF

SMALL-WARES

is complete, as also our Stock of

ZEPHYR WORSTED, SHETLAND WOOL

ALSO,

GERMANTOWN WOOL

IN ALL COLORS.

Most of our Goods being our OWN MANUFACTURE; OR, IMPORTATION, *we can offer particular inducements in prices.*

Soliciting a call, or your orders by mail,

We remain, Yours, Respectfully,

WM. H. HORSTMANN & SONS,

Fifth & Cherry Sts.

February, 1865.

Broadside: William H. Horstmann & Sons, February, 1865. Many of Horstmann's military products were sold by this firm rather than by Horstmann Brothers & Company which had been formed a few years before the Civil War to handle the military business. (Courtesy Norm Flayderman)

Dress Trimmings.

- SILK BELTINGS, WIDE & NARROW
- UNION " "
- SQUARE BUTTONS,—GILT, JET, PEARL AND SILK.
- BALL BUTTONS,—GILT, JET, SILK AND PEARL.
- GILT, STEEL & JET BUCKLES, WIDE
- " " NARROW
- BEAD AND BUGLE ORNAMENTS.
- BEAD ORNAMENTS IN SETS.
- BUGLE AND BEAD GIMPS.
- TUBULAR BRAIDS.
- SILK DRESS CORDS.
- GUIPURE LACES.
- NOTTINGHAM LACES.
- SILK CLOAK TASSELS.
- EMBROIDERING BRAIDS.
- INVISIBLE NETS.
- VELVET RIBBONS.
- SILK RIBBONS, No. 4 and 5.
- TAFFETA RIBBONS.
- ALPACA BRAIDS.
- SKIRT BRAIDS.
- CRAPE TRIMMINGS.
- CROCHET BRAIDS,—COTTON AND WORSTED.
- MAGIC RUFFLES.
- WATER PROOF CUFFS & COLLARS
- SHAWL BORDERS.

Small-Wares.

- SPOOL COTTON.
- PATENT THREAD.
- SEWING AND MACHINE SILKS.
- STAY BINDINGS.
- LINEN TAPES AND BRAIDS.
- SKIRT BRAIDS.
- CORSET AND SHOE LACETS.
- SILK AND UNION GALLOONS.
- AMERICAN AND ENGLISH PINS.
- NEEDLES.
- HOOKS AND EYES.
- COTTON CORDS.
- FRENCH WORKING COTTON.
- CROCHET AND MARKING COTTON.
- LINEN FLOSS.
- INDIA RUBBER BUTTONS.
- PEARL AND AGATE BUTTONS.
- COAT AND VEST BUTTONS.
- ELASTIC CORD AND BRAIDS.
- GARTER WEBBS AND CLASPS.
- CORSET CLASPS.
- COUTILLE & SATTEEN for CORSETS
- WHALE BONE.
- INDIA RUBBER COMBS.
- IVORY AND HORN COMBS.
- TEETH BRUSHES.
- DEXTER'S COTTON.
- INDIA MENDING COTTON.
- WOOLEN YARNS, &c.

Zephyr Department.

- BERLIN ZEPHYR WORSTED,—2, 4 and 8 Fold
- SAXONY WOOL.
- ANDALUSIAN WOOL.
- GERMANTOWN WOOL.
- CASHMERE WOOL.
- BRILLIANT ZEPHYR.
- SHETLAND WOOL.
- EMBROIDERED SLIPPERS.
- " CHAIR SEATS.
- " CUSHIONS.
- SLIPPER and other EMBROIDERING PATTERNS.
- COTTON AND LINEN CANVASS
- BLACK AND BRILLIANT "
- CROCHET NEEDLES.
- KNITTING NEEDLES.
- NETTING "
- FILLING FLOSS.
- FLOSS SILK.
- EMBROIDERING SILK.
- TATTING SHUTTLES.
- SOFT CHENILLE.
- WIRE CHENILLE.
- EMBROIDERING CHENILLE.
- STEEL AND GILT BEADS.
- GLASS BEADS AND BUGLES.
- WAX BEADS.

AND MANY OTHER ARTICLES.

This extensive product list shows Horstmann's intention to have William H. Horstmann & Sons supply the firm's civilian trade. (Courtesy: Norm Flayderman)

Columbia Manufacturing Company,
No. 223 CHESTNUT STREET.

Philadelphia, January 1, 1857.

M

We have the pleasure of informing you, that we have purchased from the Clinton Company, at Clinton, Mass., all the Power Looms, Machinery, Patent Rights and Interests of said Company connected with the manufacture of Coach Laces by Power Looms.

The Looms are now in course of removal to our Manufactory at Fifth and Cherry Streets, and we shall hereafter continue the business in this City.

There will be no interruption in the execution of orders, in consequence of the removal of the machinery.

The prices of the different descriptions of Laces, will be the same as charged by the Clinton Company, viz:

	NET PRICES.	SIX MONTHS.
WORSTED, 2 inches,	13 cts.	Per Yard.
do 2¼ do	15	do
do 2½ do	17	do
do 3 do	21	do
SILK STRIPED, 2½ inches,	21	do
do do 3 do	25	do
do BROCHE,	42½	do
do SEAMING & PASTING, striped,	4	do
do do do barred,	5	do
WORSTED do do	3½	do
EXTRA PASTING,	4	do

The very high quotations for Silk and Worsted, will prevent any deviation from these prices.

We take this opportunity to tender to you, our thanks for the many favors hitherto conferred, and to solicit the continuance of your business.

We are,

Very Respectfully Yours,

WM. H. HORSTMANN & SONS.

Broadside indicating that Horstmann had bought the textile machinery of the Clinton Company in Massachusetts in January, 1857 (Horstmann scrapbooks)

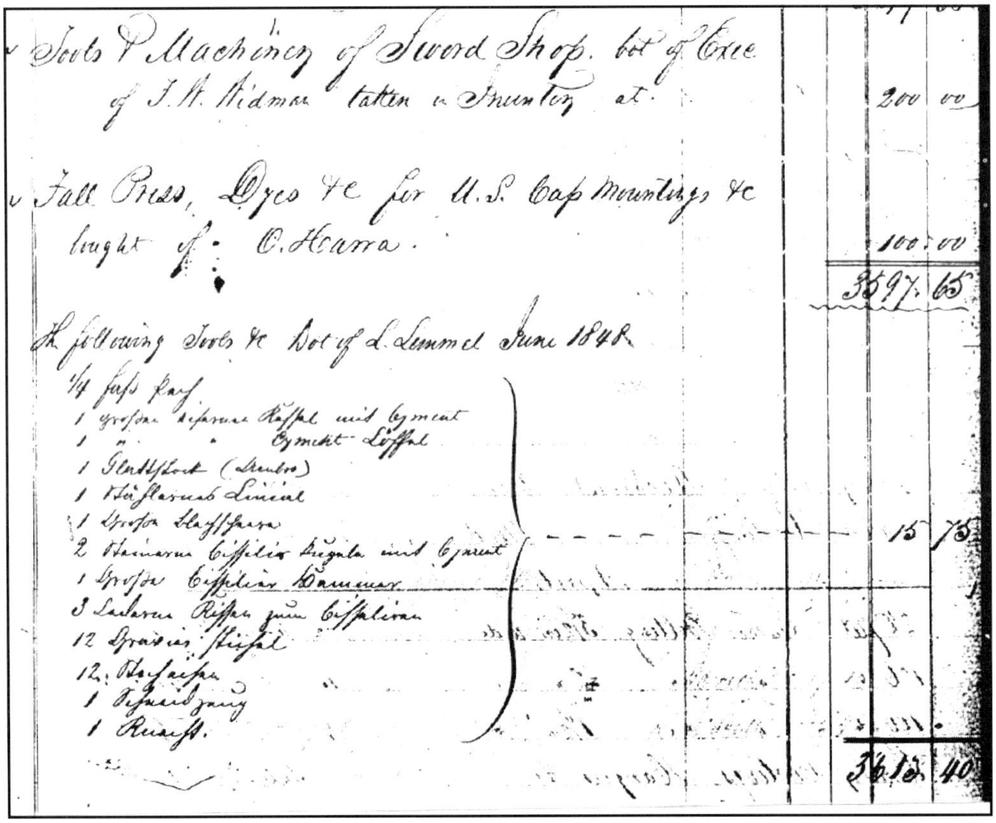

Tools & Machinery of Sword Shop. bot of Exe of J.H. Widman taken in Inventory at 200 00

Fall Press, Dyes &c for U.S. Cap Mountings &c bought of O.Harra. 100.00

3597.65

The following Tools &c Bot of L. Lemmel June 1848

15 75

3613.40

Page from the Horstmann "Machine Book" January, 1849, showing the purchase of Widmann's sword business and the insignia business of John O'Harra the previous year. Note the list of tools in German sold by Louis Loemmel to Horstmann in June, 1848. Loemmel was one of Widmann's workmen who joined Horstmann where he remained employed for at least the next 25 years. (Horstmann papers. Historical Society of Pennsylvania)

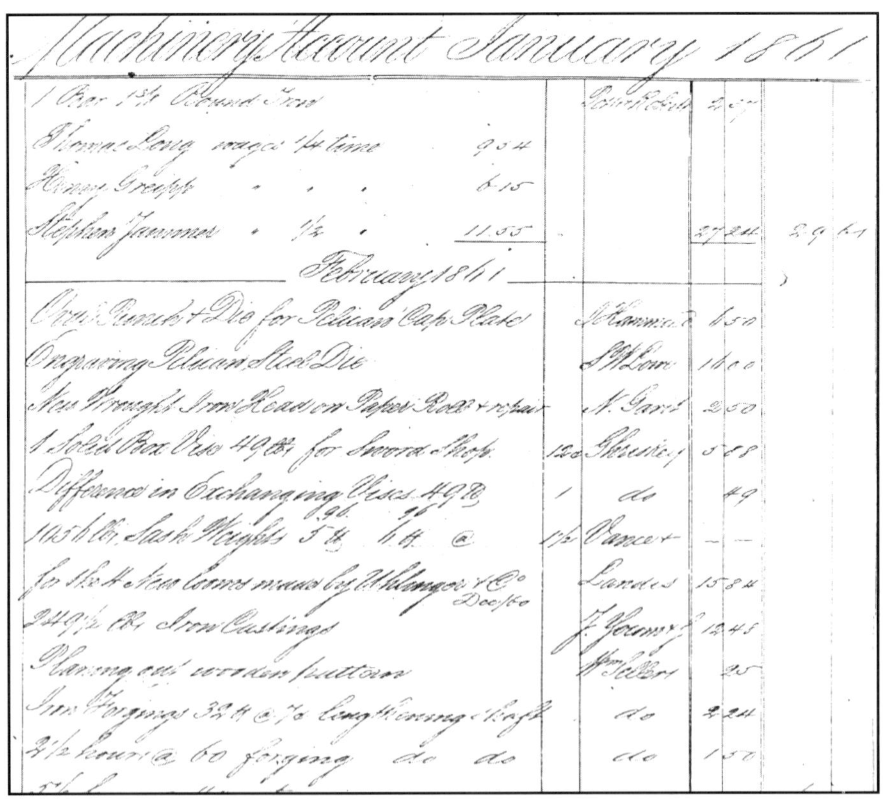

Machinery Account January 1861

February 1861

Page from Horstmann "Machine Book" dated February through April, 1861, showing dies being made for Louisiana and North Carolina buckles although those states had already left the Union (Horstmann papers)

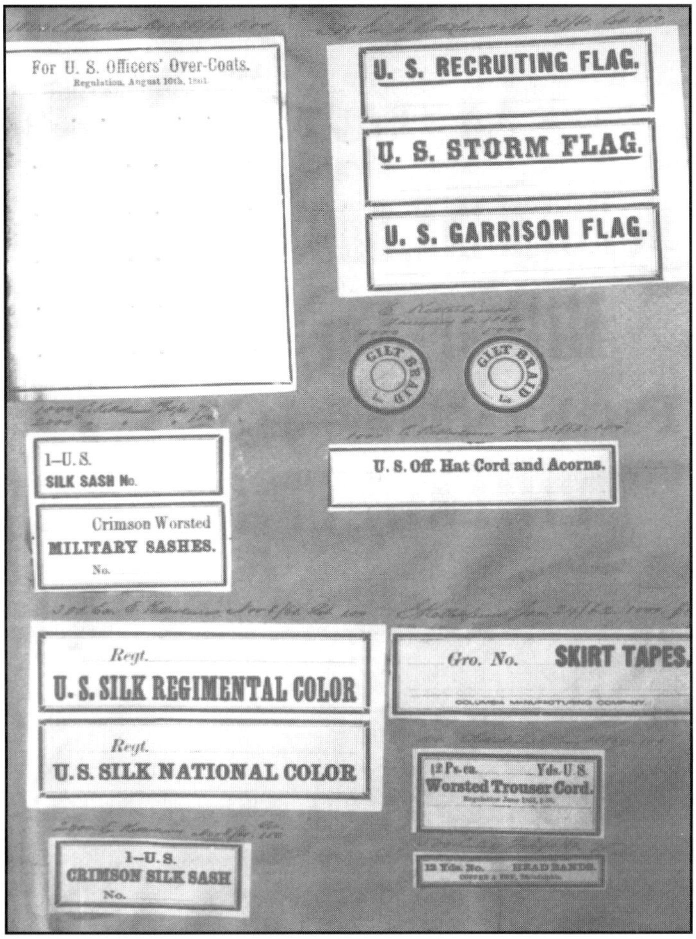

1861 labels for regimental colors and officers' overcoat buttons (Horstmann scrapbooks)

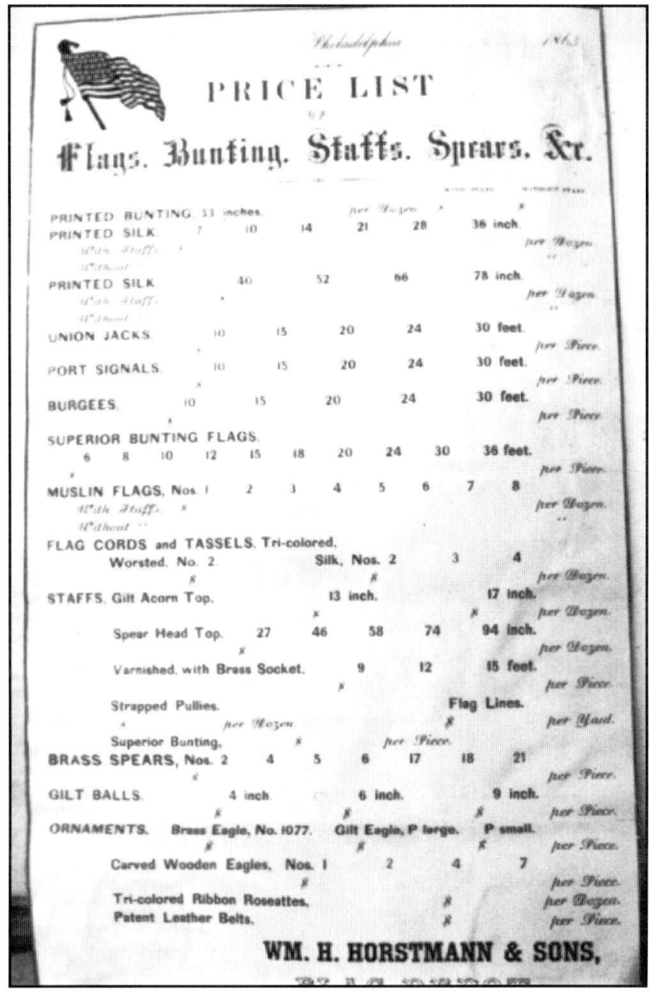

Price list for flags, 1863 (Horstmann scrapbooks)

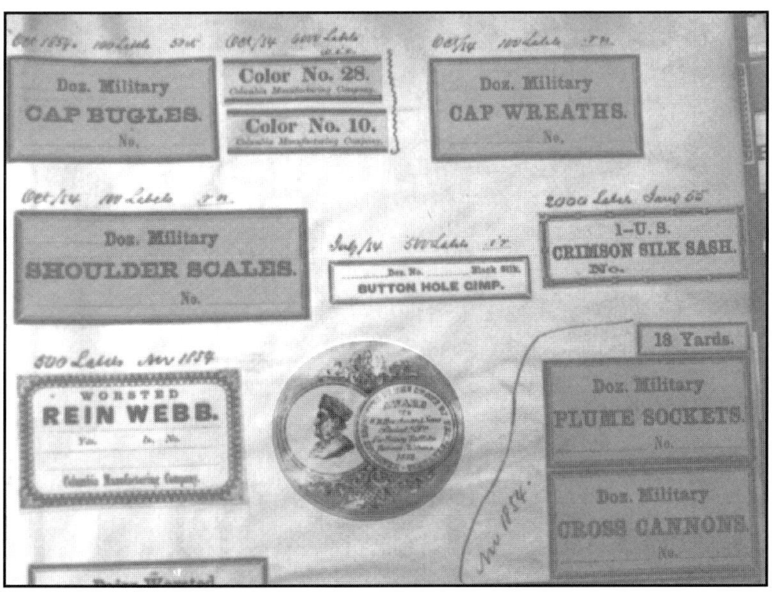

Labels showing Franklin Institute award along with Horstmann trimmings and insignia. July through November, 1854 (Horstmann scrapbooks)

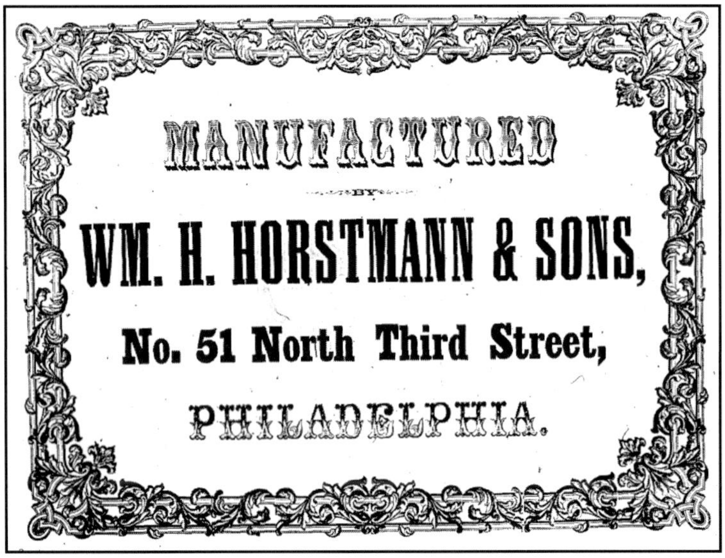

Label for the inside of Horstmann made drums. December, 1853 (Horstmann scrapbooks)

1861 label for drums and swords (Horstmann scrapbooks)

Horstmann made drum, c. 1865 (The State Museum of Pennsylvania)

Stevensgraph of Lincoln with loom instructions for its production. Made for the election of 1864. (Horstmann scrapbooks)

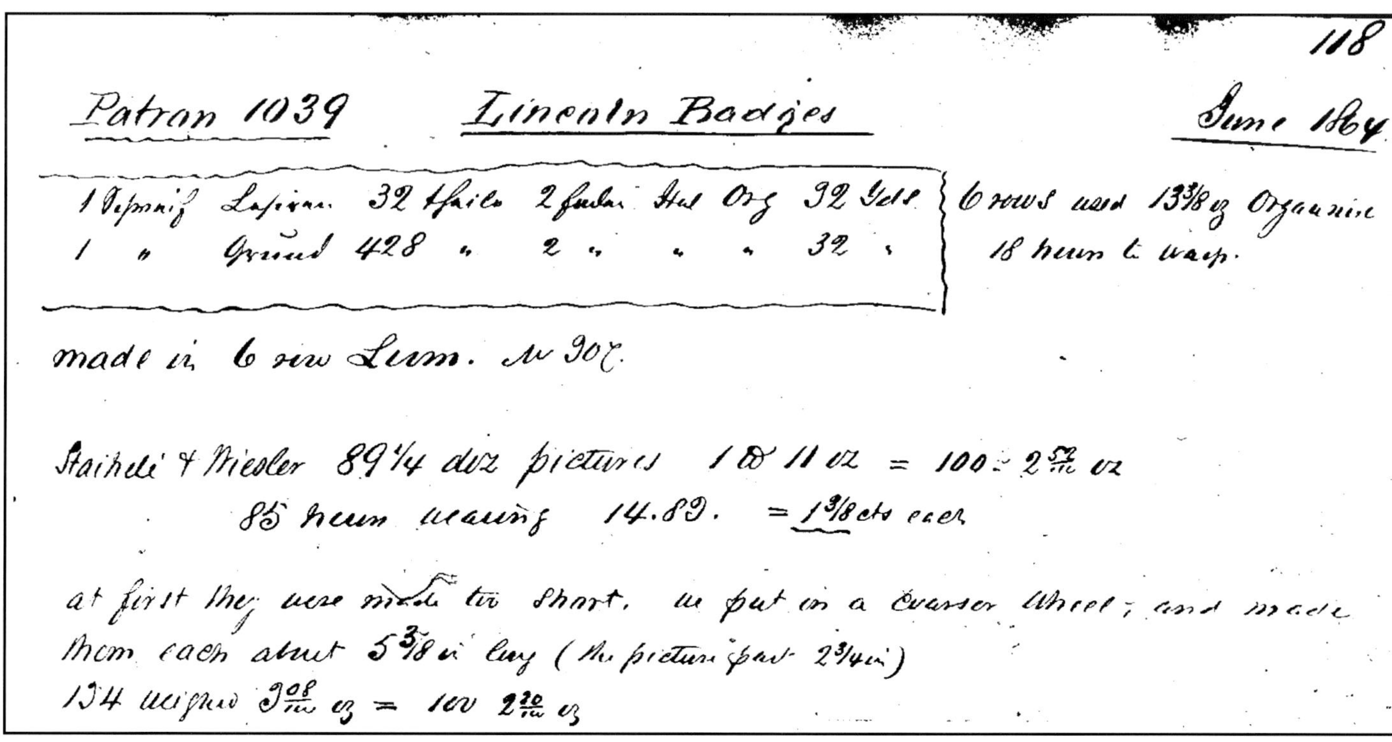

118

Patron 1039 Lincoln Badges June 1864

1 [illegible] 32 [illegible] 2 [illegible] Org 32 Yds } 6 rows used 13 3/8 oz Organzine
1 " Ground 428 " 2 " " " 32 " } 18 hours to warp.

made in 6 row Loom. No 907.

Staihele & Miesler 89 1/4 doz pictures 1 lb 11 oz = 100 - 2 [illegible] oz
85 hours weaving 14.89. = 1 3/8 cts each

at first they were made too short. we put in a coarser wheel; and made them each about 5 3/8 in long (the picture part 2 3/4 in)
134 weighed 9 08/100 oz = 100 2 30/100 oz

Die made for insignia of the Phoenix Hose Company by Horstmann. The die appears in Horstmann's machine book in August, 1854. It was cut by F. C. Key & Sons who were paid $17 for their services. (The State Museum of Pennsylvania)

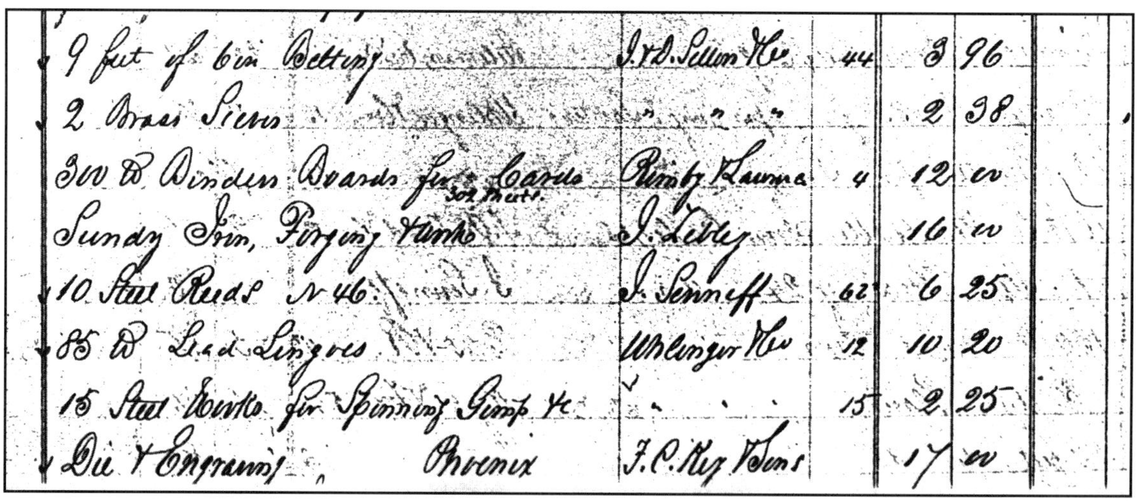

9 feet of 6 in Belting	J. & D. Sellers & Co	44	3	96
2 Brass Sieves	" "		2	38
300 lb Binders Boards for Cards	[illegible]	4	12	00
Sunday Iron, Forging [illegible]	J. [illegible]		16	00
10 Steel Reeds No 46	J. Senneff	62	6	25
85 lb Lead [illegible]	[illegible] & Co	12	10	20
15 Steel [illegible] for Spinning Gimp &c	"	15	2	25
Die & Engraving Phoenix	F. C. Key & Sons		17	00

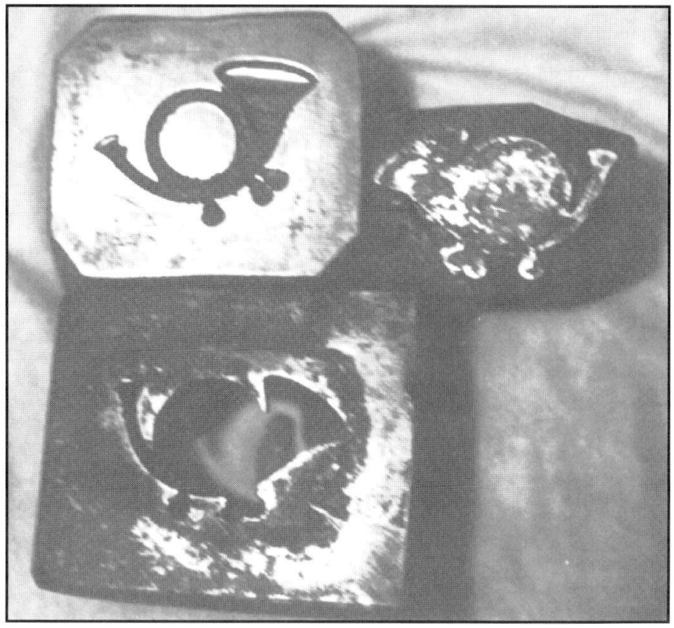

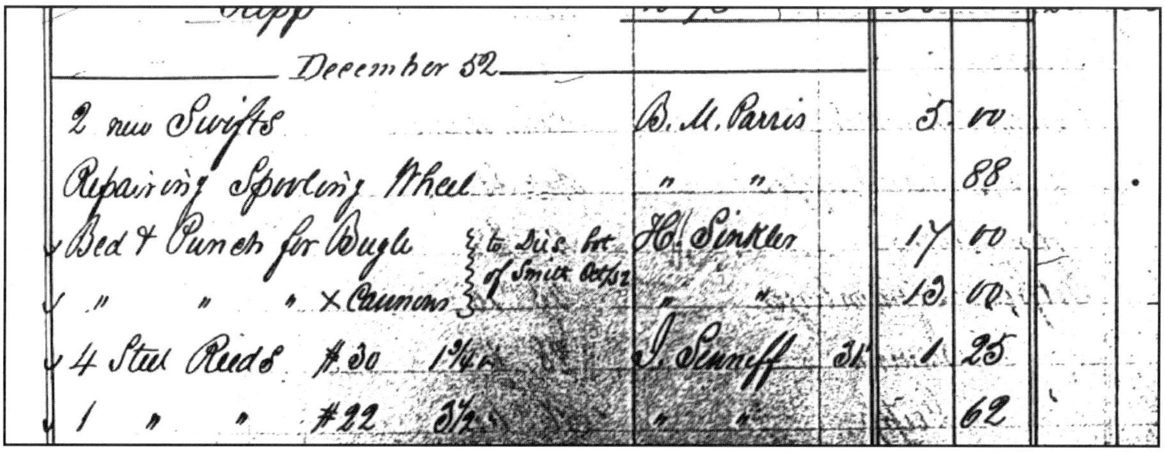

December 52

2 new Swifts	B. M. Parris		5.00
Repairing Spooling Wheel	" "		88
Bed & Punch for Bugle } to Die for of Smith Oct/52	H. Sinkler		17.00
" " " × Cannons	" "		13.00
4 Steel Reeds #30 1¼	J. Senneff	31	1.25
1 " " #22 3½	" "		62

Die for M1851 officers' cap insignia. Probably reflects the machine book entry for October and December, 1852, ordering the die and the trimming tools for this insignia. (Author's collection)

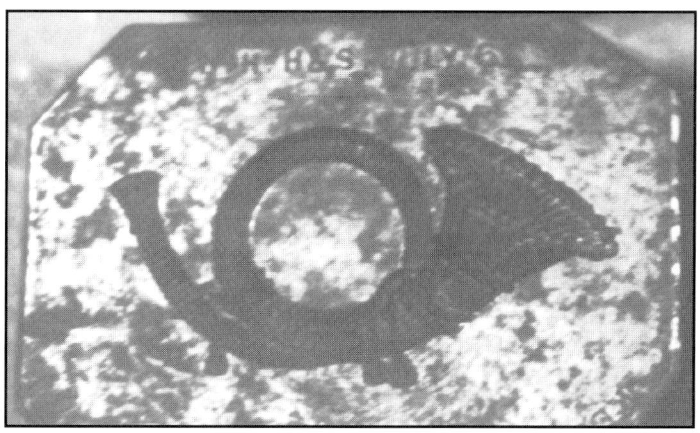

Die for M1851 infantry officers' cap insignia. Marked "W. H. H. & S. July 61". Made by F. C. Key & Sons. (Stokes Kirk collection)

M1821 eagle cap insignia. Die is marked "Leonard" and dates c. 1840. This die came through Horstmann's hands with either the O'Harra purchase of 1848 or the Pinchin purchase of 1862. The die then continued to Thomas Parry & Sons who did much of Horstmann's striking work in the late 19th century and from there to W. Stokes Kirk & Co. (Author's collection)

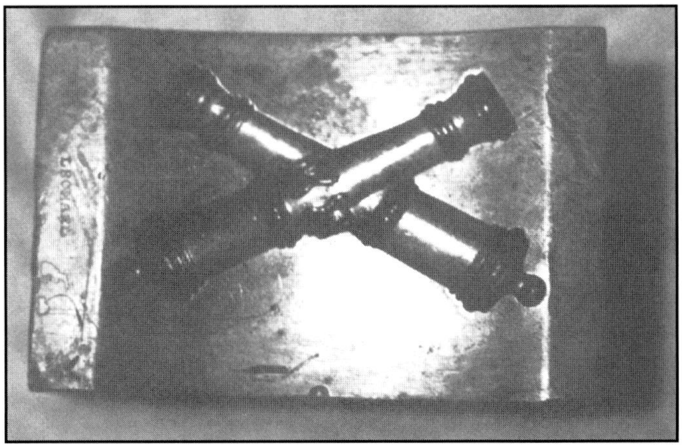

Die for M1832 artillery insignia. Also marked "Leonard" this has the same history as the plate above. (Author's collection)

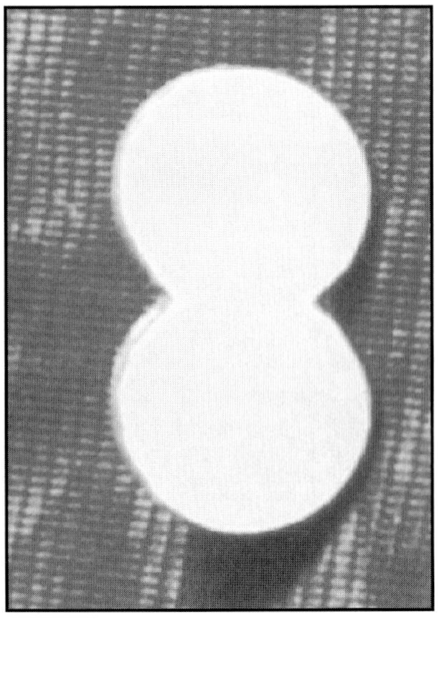

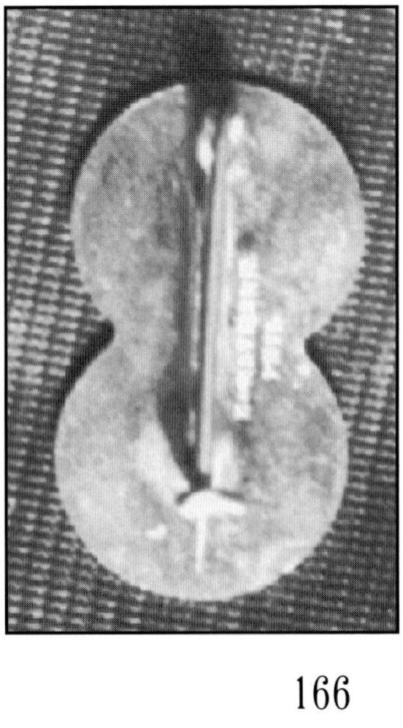

8th Corps Badge, c. 1898. Made by or for Horstmann. (Author's collection)

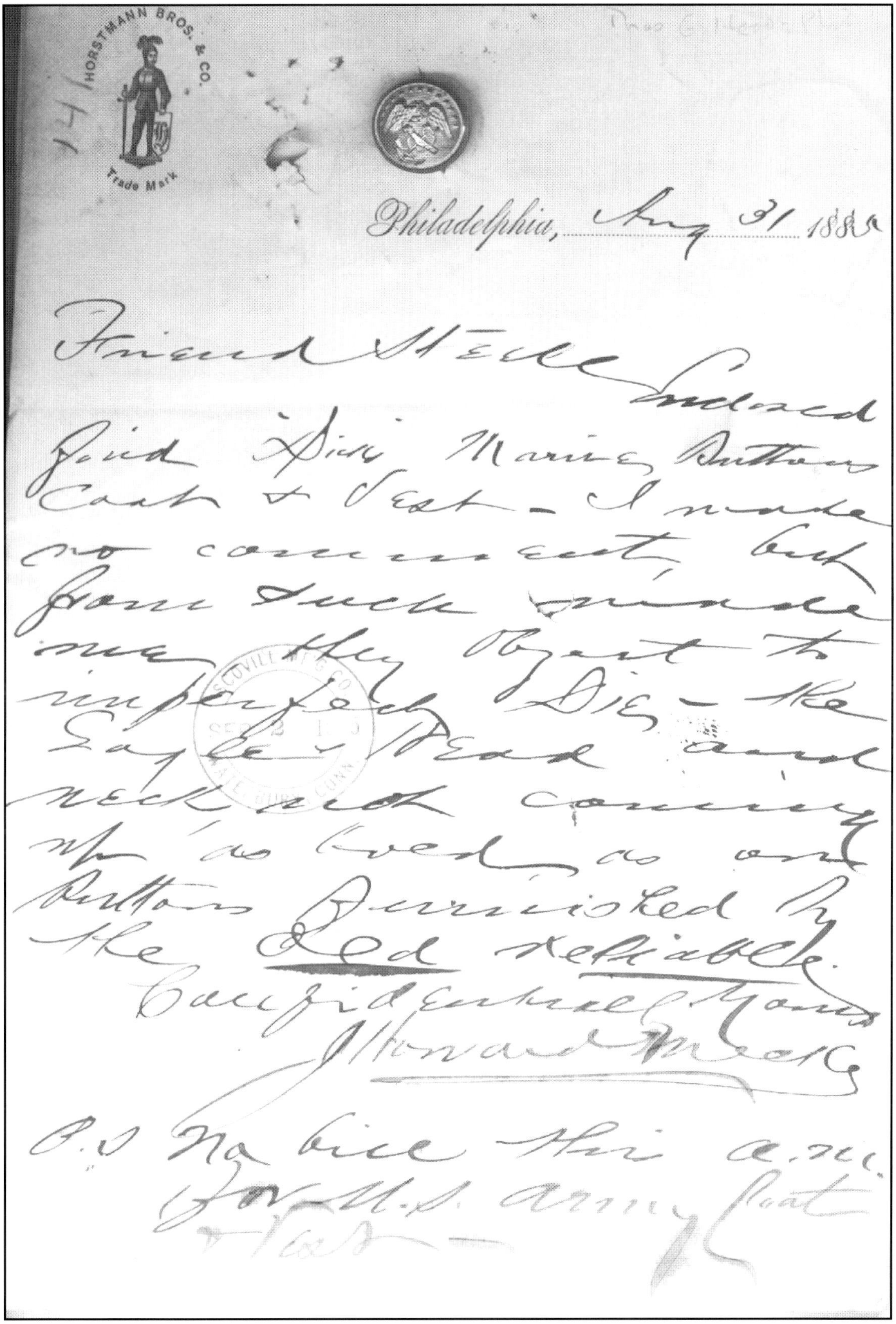

HORSTMANN BROS. & CO.
Trade Mark

Philadelphia, Aug 31 1885

Friend Steele
Enclosed
find Dick Marine Buttons
Coat & Vest - I made
no comment, but
from such minds
they object to
imperfect Die - the
Eagle's head and
neck not coming
up as bold as our
buttons furnished by
the Old Reliable.
Confidentially Yours
J Howard Mears

P.S. No bill this a.m.
for U.S. Army Coat
& Vest —

Letter from Scovill files. This letter shows Horstmann negotiating with Scovill to make buttons in 1885. The actual button attached to the 1885 letter is back marked Thos. G. Hood, a customer of Horstmann's.

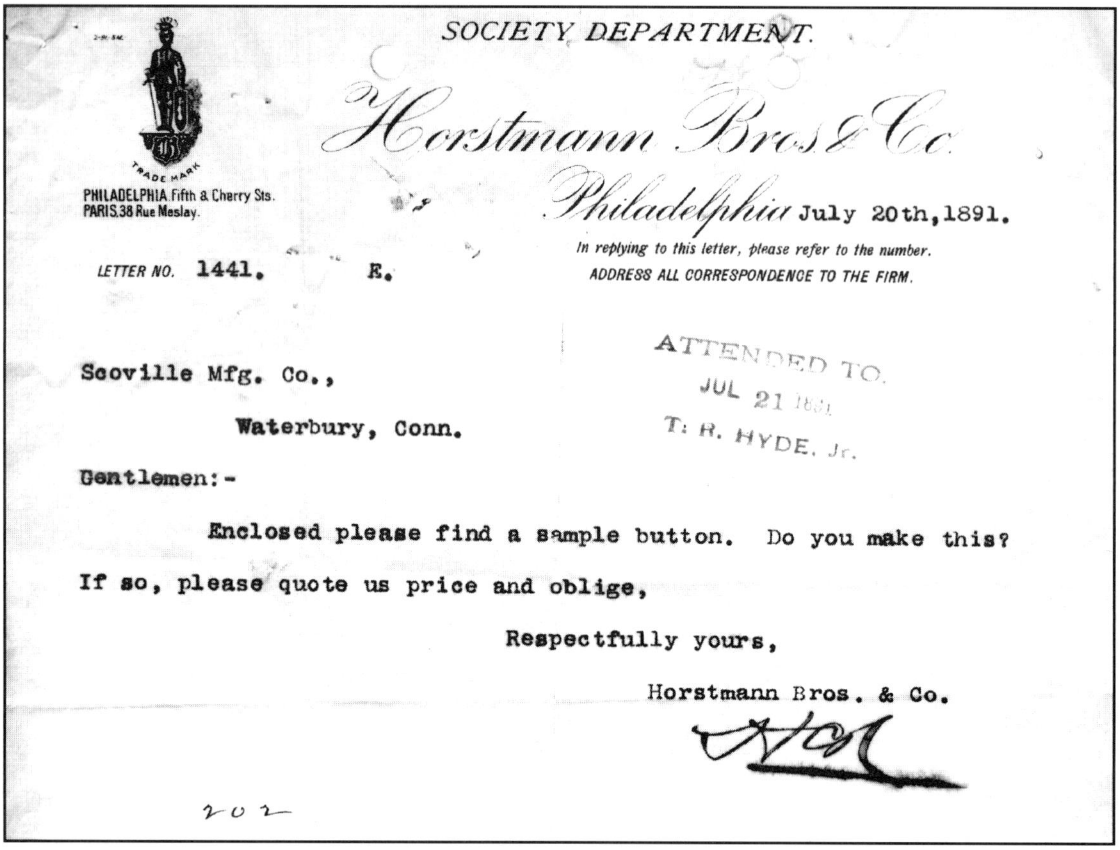

SOCIETY DEPARTMENT.

Horstmann Bros & Co.

TRADE MARK

PHILADELPHIA Fifth & Cherry Sts.
PARIS 38 Rue Meslay.

Philadelphia July 20th, 1891.

In replying to this letter, please refer to the number.
ADDRESS ALL CORRESPONDENCE TO THE FIRM.

LETTER NO. 1441. E.

ATTENDED TO.
JUL 21 1891
T. R. HYDE, Jr.

Scoville Mfg. Co.,

Waterbury, Conn.

Gentlemen:-

Enclosed please find a sample button. Do you make this? If so, please quote us price and oblige,

Respectfully yours,

Horstmann Bros. & Co.

202

Letter from Scovill files. This letter shows Horstmann negotiating with Scovill to make buttons in 1891.

Sample buttons for the Revenue Cutter Service marked Horstmann Bros. & Co. They appear on a Horstmann trade card but were in Scovill's sample files.

The 20th Century

"The Business Today" taken from One Hundred Years shows the Horstmann complex at 5th and Cherry in 1916 (pp. 35-54)

Administrative section

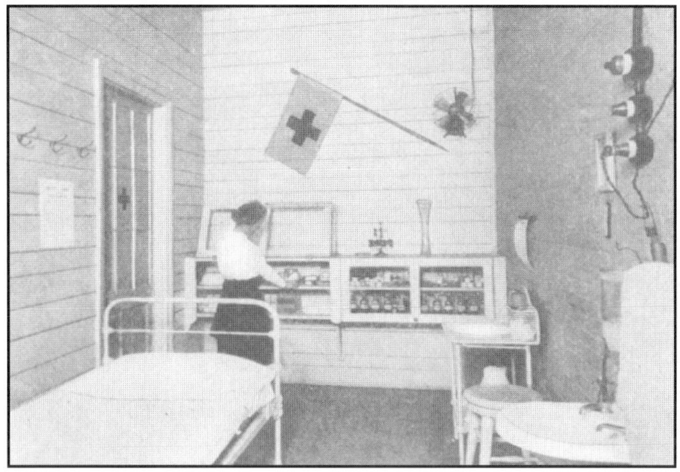

First Aid

A noonday gathering in the Recreation Room

The Lunchroom

Main Sales Rooms, Section A

Office of Yarn Department

Sales Departments, Section B

Silk throwing

Making upholstery trimmings

Putting up Columbia cottons

Braiding machines

Balling Columbia cottons

Limousine trimmings

Making flags

Jacquard ribbon loom

Sword and metal workers

Binding loom

Making stamped goods

Reeling and packing
Columbia yarns

Woodworking shop

Manufacturing regalia

Hand embroidery department

Accounting department

Horstmann advertisement removed from a trunk lid (McGuinn collection)

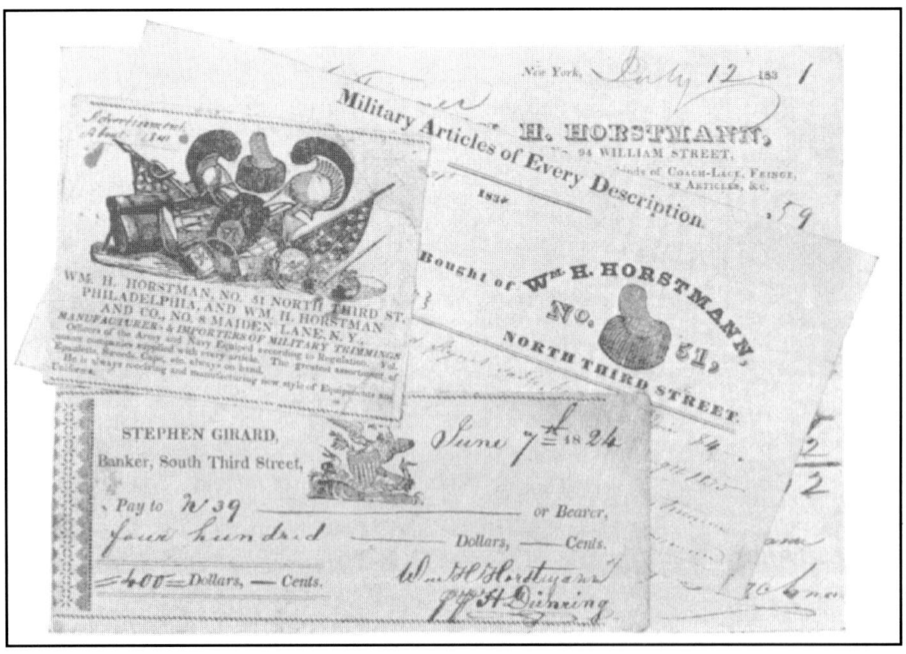

The business association of Henry Duhring and William H. Horstmann is shown from this check signed by both (*One Hundred Years*, p. 15)

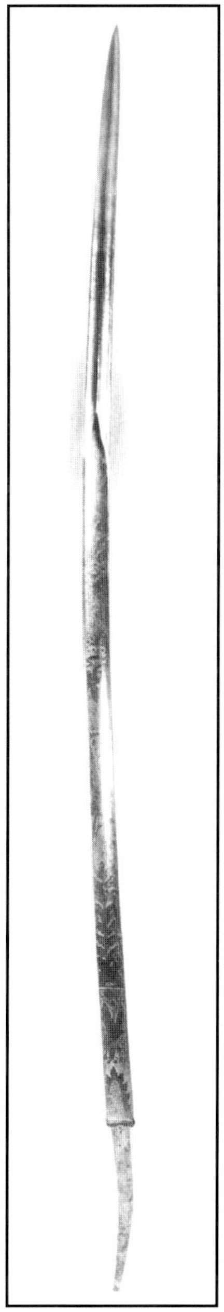

High quality blade made for F.W. Widmann by the Weyersberg family in Solingen. The Weyersberg trade mark used c. 1787-1883 was a crowned head. The initials "N.K." also appear on the tang of the blade and are probably those of the actual cutler in the Weyersberg factory. Most of these marks are covered when the blade is hilted but a series of Civil War period Horstmann blades has been noted which are similarly unmounted. These also have the crowned head Weyersberg mark and "N.K." but several also have the initials "C.V." and "Damascus" as well. It would seem that in addition to Widmann's tools and workers, Horstmann also used Widmann's established business with Weyersberg to obtain at least some of their blades. (Author's collection)

REGULATIONS

FOR THE

UNIFORM & DRESS

OF THE

ARMY OF THE UNITED STATES.

JUNE, 1851.

From the original Text and Drawings in the War Department.

SECOND EDITION.

PHILADELPHIA:

WILLIAM H. HORSTMANN & SONS,

MILITARY FURNISHERS.

LIPPINCOTT, GRAMBO & COMPANY.

MDCCCLI.

Cover of book offering new Army uniforms conforming to the 1851 regulations. The signature of Adjutant General Jones shows that the book was intended to serve as an official publication (see pages 177-178). (Courtesy: Norm Flayderman)

Official
R Jones. Adjut Genl
Washington June 1857.
Captain
Foot Riflemen.
Ordnance Department.

Official
R Jones. Adjut Genl
Washington June 1857
Major
Adjt General's Dep
Captain.
Engineers.
Engineer Soldiers.

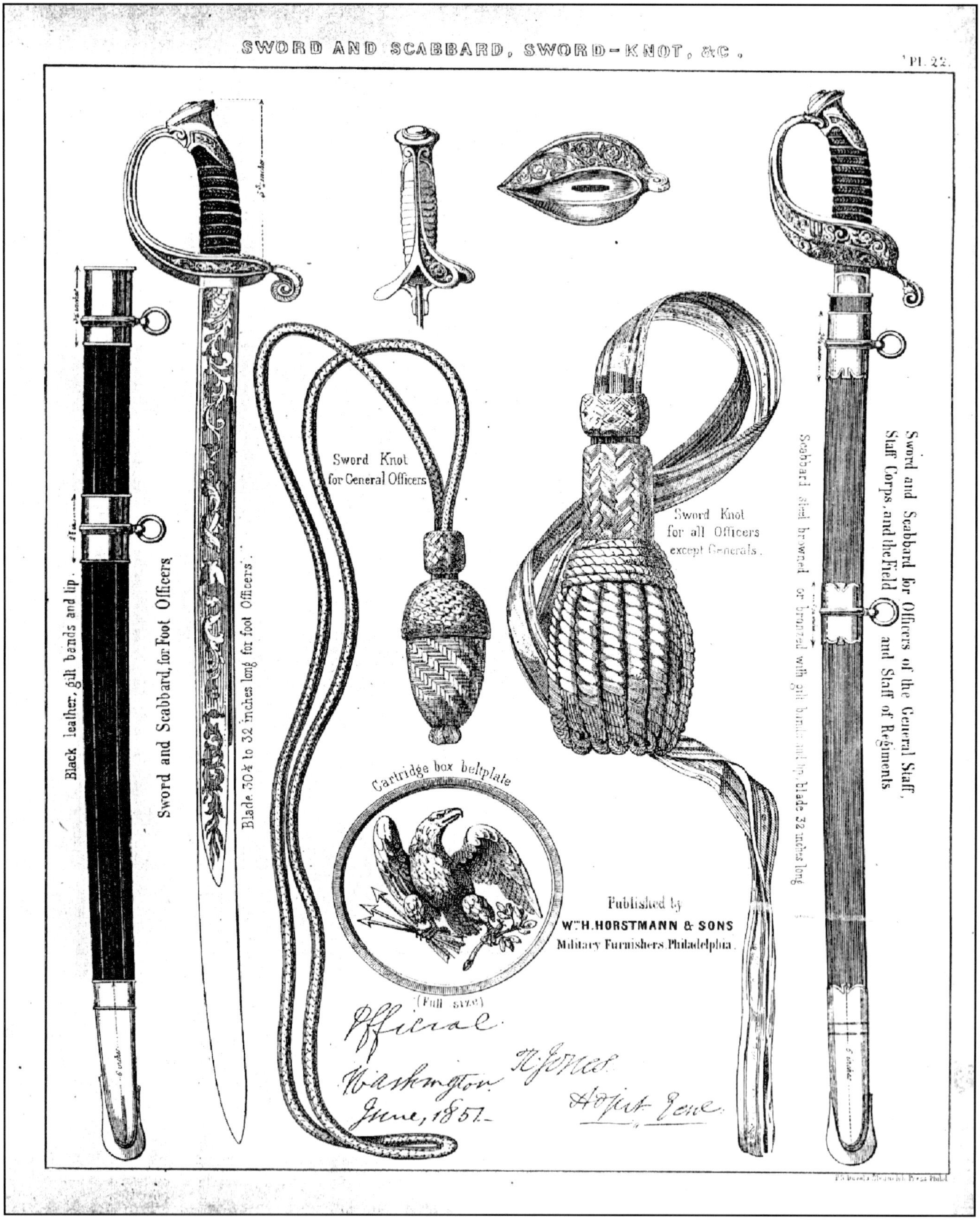

SWORD AND SCABBARD, SWORD-KNOT, &C.
PL. 22.
Black leather, gilt bands and tip.
Sword and Scabbard, for Foot Officers
Blade 30¾ to 32 inches long for foot Officers.
Sword Knot for General Officers
Sword Knot for all Officers except Generals.
Sword and Scabbard for Officers of the General Staff, Staff Corps, and the Field and Staff of Regiments
Scabbard steel browned or bronzed with gilt bands and tip, blade 32 inches long
Cartridge box beltplate
(Full size)
Published by
Wm H. HORSTMANN & SONS
Military Furnishers, Philadelphia.
Official
Washington
June, 1851
R Jones
Adjt Genl

Horstmann exhibit in 1876 U. S. Centennial held in Philadelphia (Horstmann scrapbooks)

Overview of the Centennial with arrow denoting the position of the Horstmann booth (Horstmann scrapbooks)

Stevensgraph of Horstmann factory at 5th and Cherry, c. 1865 (Horstmann scrapbooks)

Print: Charles Oakford and Sons store interior, c. 1860. The interior is typical of businesses of the era. (Courtesy: The Library Company of Philadelphia)

Chapeau de bras and box label of Commodore James Biddle. Sold by Horstmann at 51 N. 3rd, probably shortly before Biddle's death in 1848. (Courtesy: Andalusia Museum)

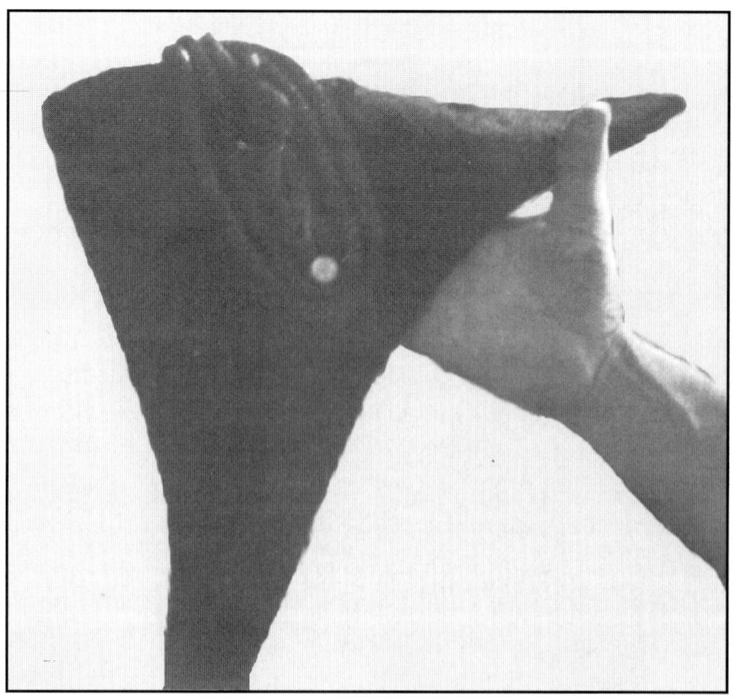

M1851 dress shako sold by William H. Horstmann & Sons at 51 N. 3rd Street, c. 1851-55. While Horstmann probably did not actually make the cap itself, the woven insignia were within the firm's manufacturing specialty. (Private collection)

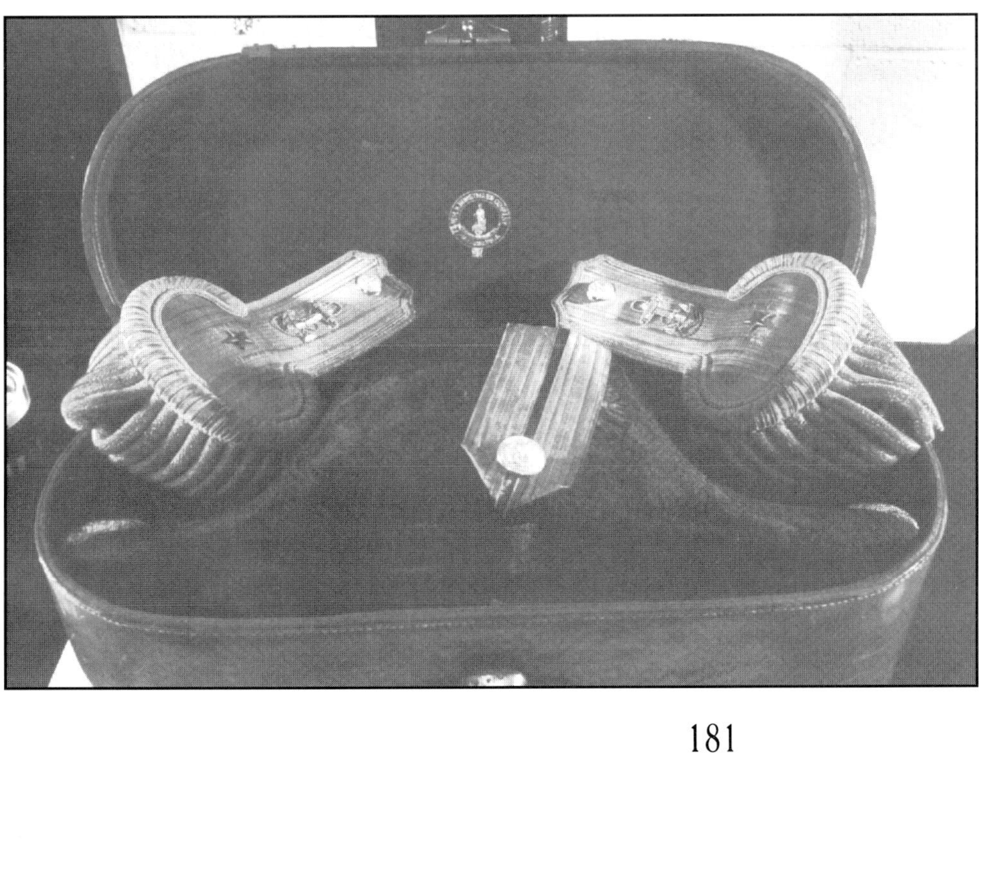

Dress chapeau and epaulettes of Admiral Fahrney, c. 1920. Sold by The William H. Horstmann Company. (Courtesy: The State Museum of Pennsylvania)

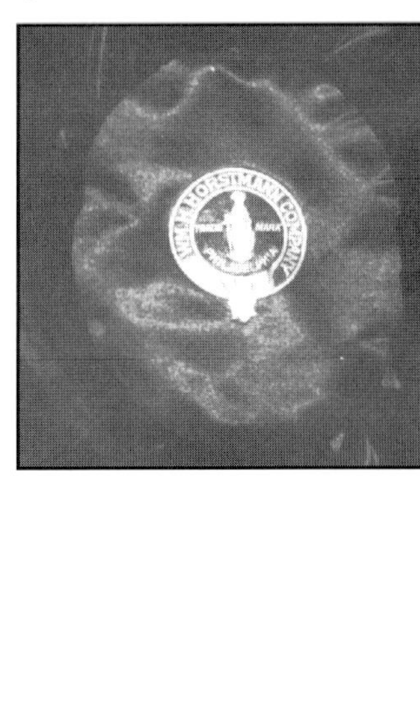

Chapeau and dress cap of LTC Seltzer, Aide to the Governor of Pennsylvania, c. 1902. Both labeled: The William H. Horstmann Company. Philadelphia. (Courtesy: The State Museum of Pennsylvania)

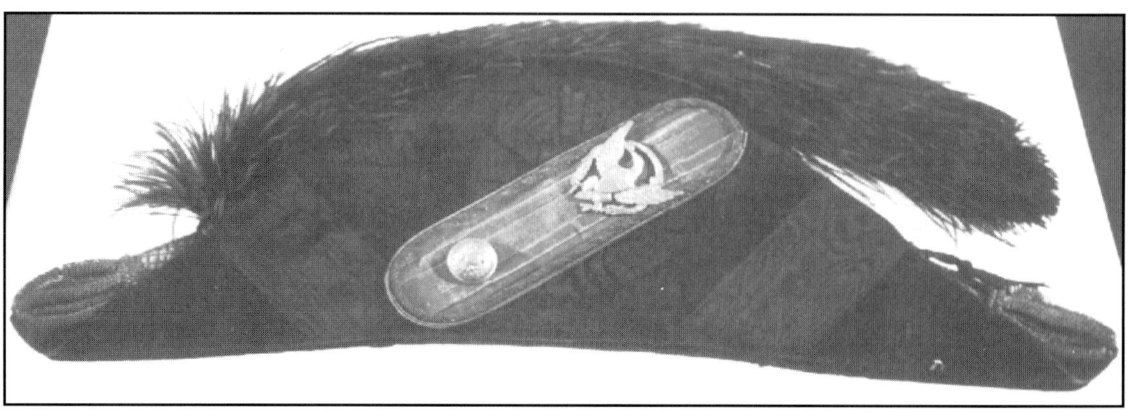

Button die: State Fencibles c. 1880. The Fencibles were a well known Philadelphia militia unit. The die is marked Horstmannn but has a Scovill tool number 2817. (Scovill die collection. The State Museum of Philadelphia)

Dies for shooting badges, worn as collar insignia. Marked Horstmann Bros January 1880. (Scovill die collection. The State Museum of Pennsylvania)

Die for "Schuylkill Navy" buttons marked "Horstmann Phil 1876". The die was cut by Jarvis E. Ellis, Scovill's master die cutter. The Schuylkill Navy was a quasi-military social organization in Philadelphia. (Scovill die collection. The State Museum of Pennsylvania)

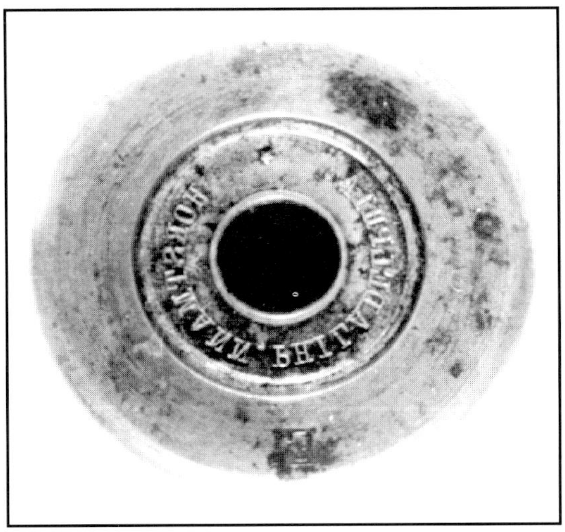

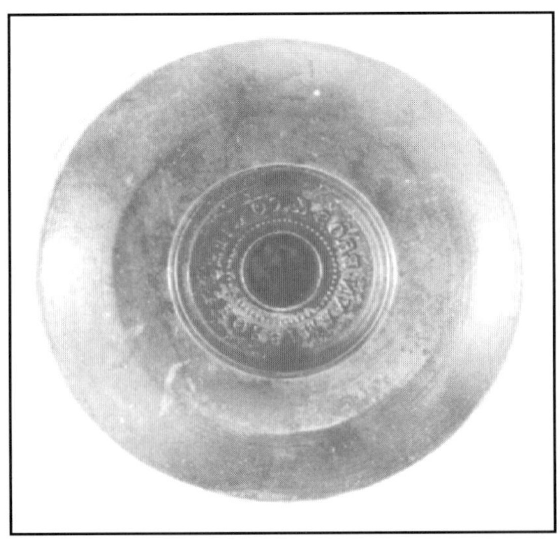

Master hob and working dies for making backs of buttons with Scovill customer name, in this case Horstmann / Philadelphia and Horstmann Bros. & Company. (Scovill die collection. The State Museum of Pennsylvania)

1840-45 (Before W.H.H. & Sons) Die used through mid-1850's

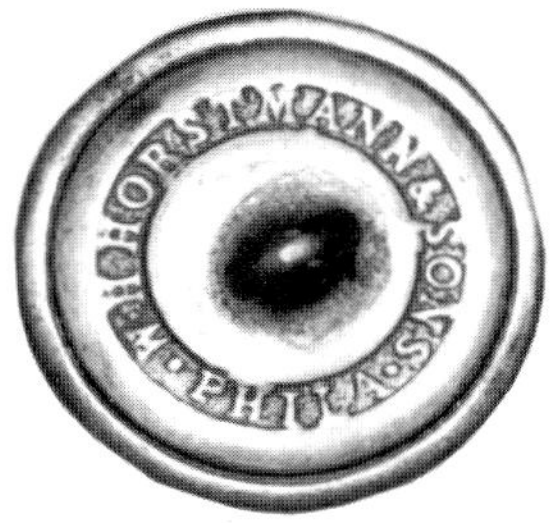

1845-60 (Post W.H.H. & Sons)

1850's

1850's

1859-70

1860-70

1860-70

c. 1860-70

1870-93

c. 1870-93

1870-93

1875-93

1870-93

1880-93

1880-93

1880-93

1894-1923

1865-76

1865-76

1894-1923

1850-65

1850-65

1850-65

1860-65

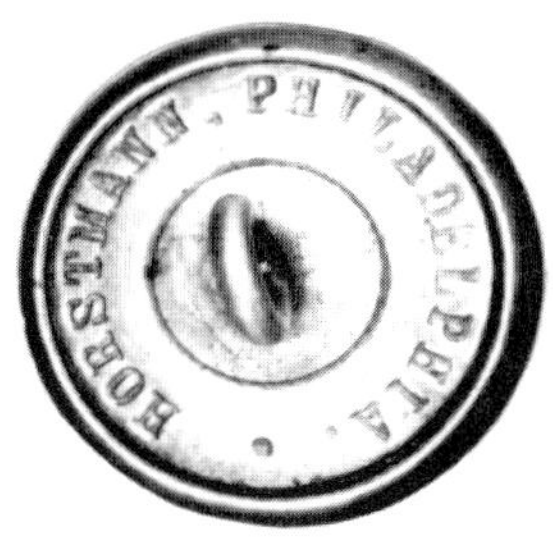

1893-1947

1865-76
Alabama State Seal
with this b/m
may be pre-CW front
with post-War back

c. 1870-76

1845-60

c. 1840-45

c. 1845-50

c. 1850-65

c. 1845-50

Militia Artillery
c. 1900 made
in 1812 style

1840-45

Horstmann never made a button. Most Horstmann buttons were subcontracted to the Waterbury firms of Lane Mfg. Co., Scovill, and Waterbury Button Co.

CivilWar
General Staff

1894-1940

1984-1940

Civil War Navy button
ordered from Waterbury

U.S. Army
1872-1902

U.S. Artillery
1835-40

U.S. Marines
Civil War period

Grand Army
of the Republic
1868-1900

Militia ?
to Morris Guard

1845-60

1894-1900

W.H. Horstmann
Phila one-piece
c.1835-50

HORSTMANN BROTHERS & CO., PHILADELPHIA. 69

STATE BUTTONS.

Maine. New Hampshire. Vermont. Massachusetts.

Rhode Island. Connecticut. New York. New Jersey.

Pennsylvania. Maryland. Virginia. North Carolina.

South Carolina. Georgia. Alabama. Mississippi.

Louisiana. Texas. Arkansas. Kentucky.

Ohio. Missouri. Michigan. Wisconsin.

Page 69 from the 1877 Horstmann catalog showing state buttons. Like Horstmanns' other products, the buttons sold show a mixture of earlier and contemporary products.

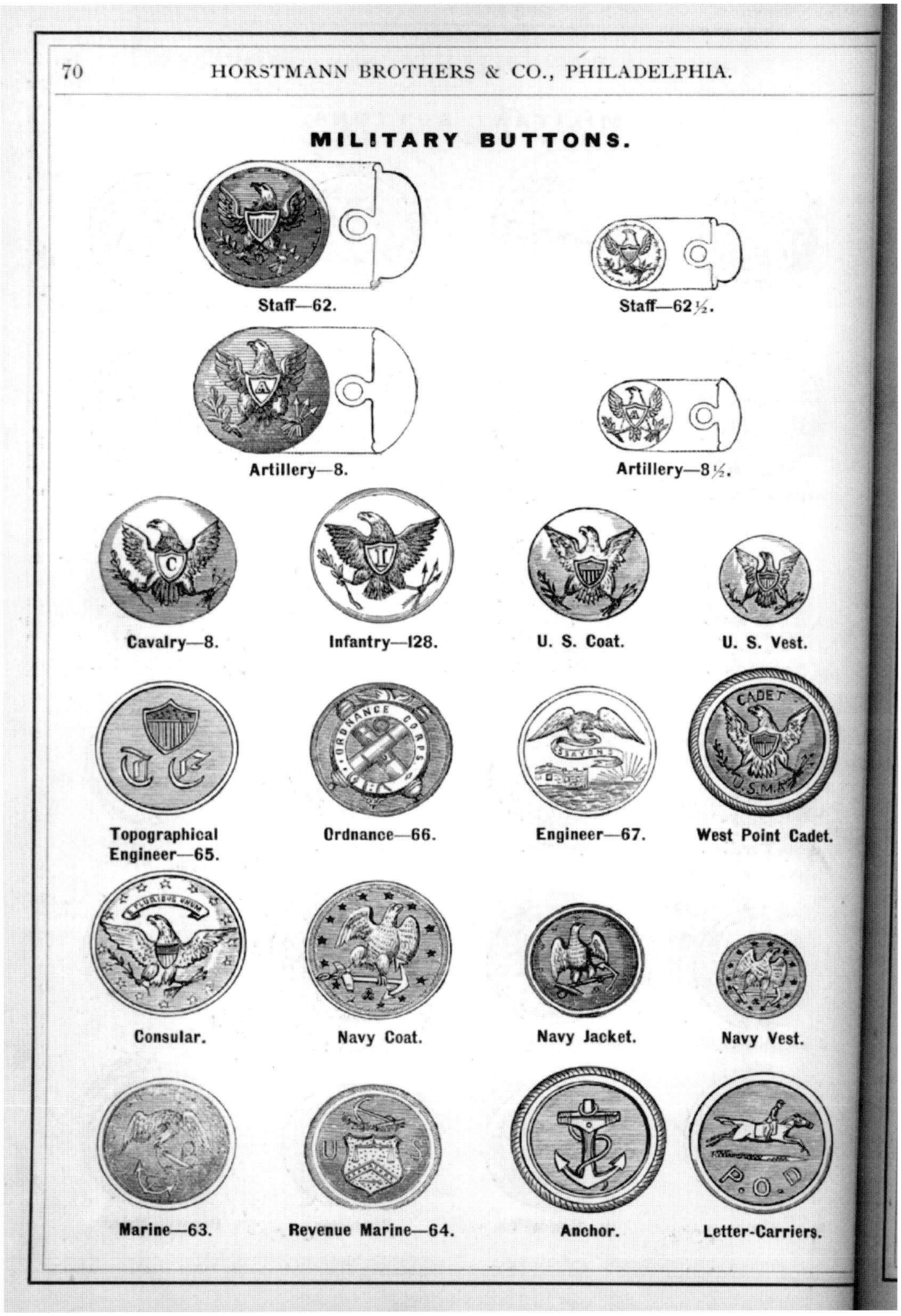
70 HORSTMANN BROTHERS & CO., PHILADELPHIA.

MILITARY BUTTONS.

Staff—62.
Staff—62½.
Artillery—8.
Artillery—8½.
Cavalry—8.
Infantry—128.
U. S. Coat.
U. S. Vest.
Topographical Engineer—65.
Ordnance—66.
Engineer—67.
West Point Cadet.
Consular.
Navy Coat.
Navy Jacket.
Navy Vest.
Marine—63.
Revenue Marine—64.
Anchor.
Letter-Carriers.

1877 catalog depiction of buttons. Many of the types shown are typical of the Civil War and the Topographical Engineer button was totally obsolete after 1862.

ESTABLISHED 1816.
INCORPORATED 1893.

TRADE MARK REG. U.S. PAT. OFF.

PHILADELPHIA, Fifth & Cherry Sts.
NEW YORK, Broadway & Grand St.
BOSTON, 7 Temple Place.
PITTSBURG, 126 Sixth Street
LYONS, Quai de Retz 10.
PARIS, 38 Rue Meslay.
LONDON, 71 & 72 Little Britain E.C.
BERLIN, Alexandrinen Str. 97
VIENNA VI Windmühlgasse 55.

In replying to this letter please refer to number
ADDRESS ALL CORRESPONDENCE TO THE COMPANY.

Military Uniforms and Equipments for
ARMY AND NAVY
NATIONAL GUARD
MILITARY ORGANIZATIONS
POLICE AND FIRE DEPARTMENTS
LETTER CARRIERS
RAILROADS
STEAMSHIP COMPANIES
SCHOOLS, GRAND ARMY
SONS OF VETERANS

Regalia, Jewels, Costumes and Secret Paraphernalia Hall Furnishings Badges Banners and Caps and all Parade Supplies for Lodges and Secret Societies

MASONIC–BLUE LODGE
CHAPTER, R. A. M.
COUNCIL R. & S. M.
KNIGHTS TEMPLAR
A. A. SCOTTISH RITE
ORDER EASTERN STAR
MYSTIC SHRINE
ODD FELLOWS–LODGE
ENCAMPMENT, CANTON, P. M.
REBEKAH DEGREE
KNIGHTS OF PYTHIAS
LODGE, UNIFORM RANK
RATHBONE SISTERS
RED MEN
AMERICAN MECHANICS
KNIGHTS OF MALTA
K. O. T. M., B. P. O. ELKS
SELECT KNIGHTS, A. O. U. W.
P. O. S. OF A., A. O. F.
KNIGHTS GOLDEN EAGLE
AND ALL SECRET ORDERS

Gold and Silver Trimmings and Costumer's Materials Flags, Banners, Bunting Fencing Supplies
Catalogs Free

Wm. H. Horstmann Company,
Fifth & Cherry Sts.
2492 F. H. T. *Philadelphia,* April 2nd, 1909.

Mr. Geo. W. Wagenseller,
c/o The Middleburgh Post,
Middleburgh, Pa.

Dear Sir:-

In reply to your esteemed favor of recent date, would state that we are mailing you herewith a revised list of Chapter Paraphernalia, which corresponds with the new regulation described by the Grand Chapter of the State of Pennsylvania, of which you have, no doubt, a book of the Constitution explaining thoroughly every article; if not, you can write the Grand Secretary, Masonic Temple, Philadelphia, and he will be pleased to mail you one. The articles listed herein are guaranteed in every respect to be strictly in accordance with the Pennsylvania regulation.

When you are ready to purchase, if you could have your Committee call, or if they wish the writer would call on the Committee and we could more readily explain things so that it will be more satisfactory to you. We have some samples of the different articles here but not of the newer articles as we

Correspondence from 1909 showing Horstmann quoting Masonic fraternal costume. Note the medals which form the letterhead as Horstmanns' 19th century product awards. (Author's collection)

The Horstmann trademarks. During the last quarter of the 19th century the knight in armor, reminiscent of a Solingen sword blade mark, came to stand for the military business of Horstmann Bros. & Company. Similarly the figure of Columbia as a seated shepherdess was used to symbolize the civilian yarn and cloth business. The use of Columbia was probably intentionally reminiscent of the Columbia Manufactory as Horstmann's early factory. These symbols first appear in Horstmann's label sample books in December, 1871, and March, 1874, respectively, as fabric labels. The label incorporating the standing knight is on a box of U. S. Marine Corps buttons c. 1900. (Author's collection)

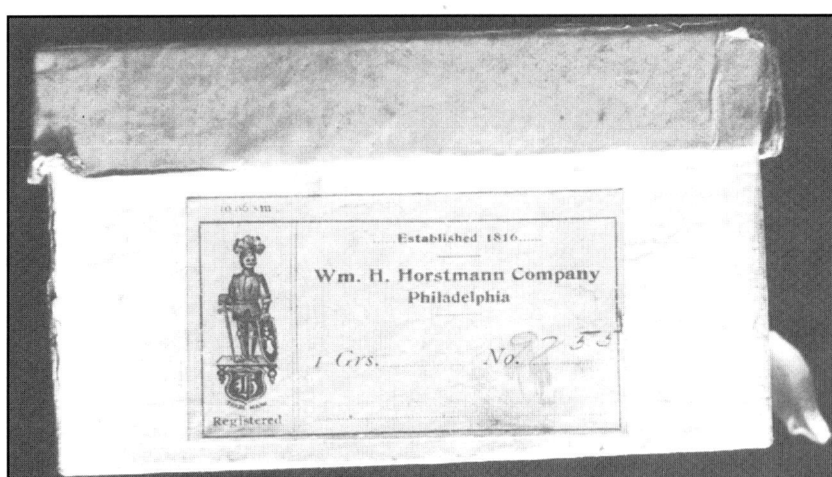

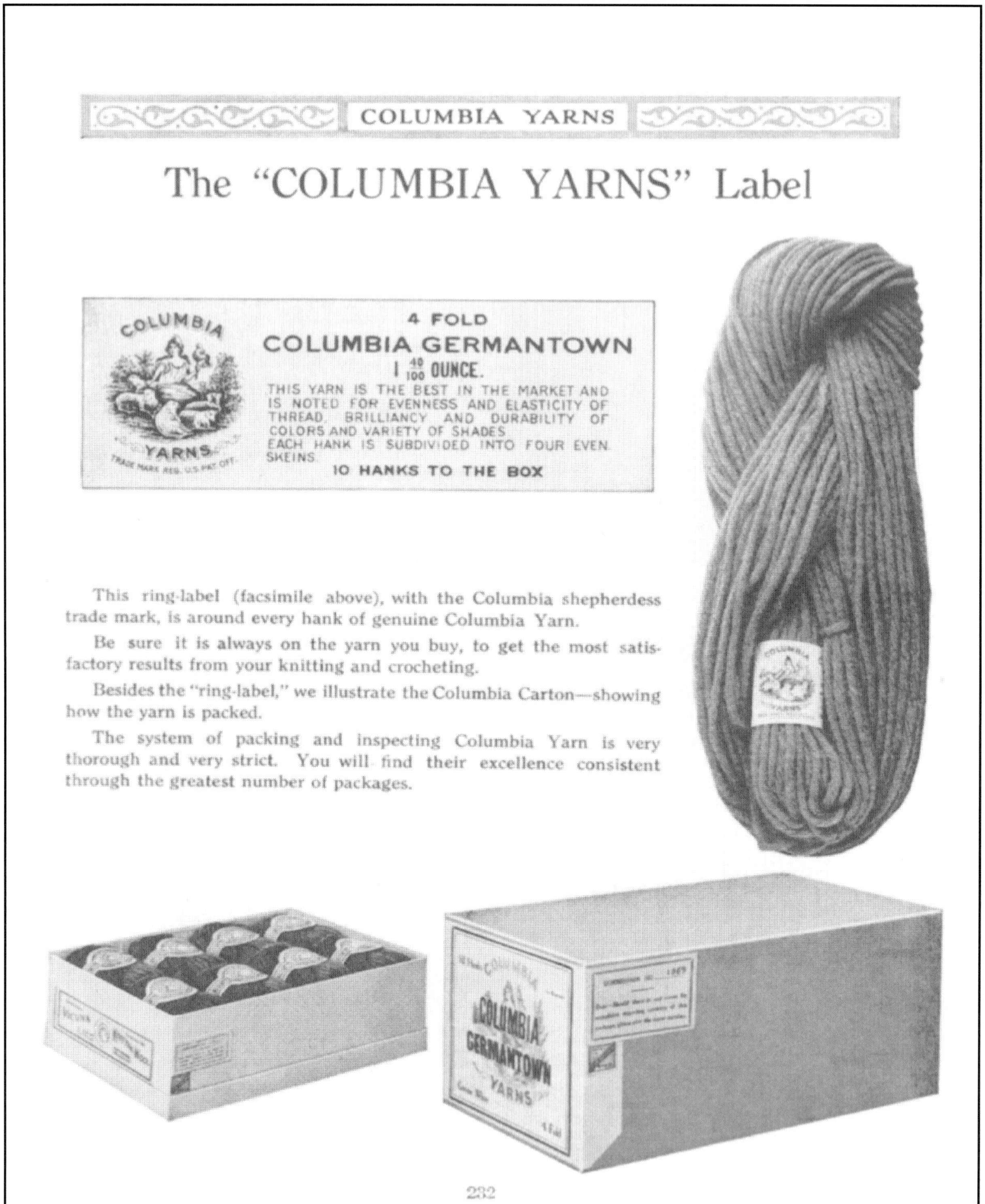
COLUMBIA YARNS

The "COLUMBIA YARNS" Label

COLUMBIA YARNS TRADE MARK REG. U.S. PAT. OFF.

4 FOLD
COLUMBIA GERMANTOWN
1 $\frac{40}{100}$ OUNCE.
THIS YARN IS THE BEST IN THE MARKET AND IS NOTED FOR EVENNESS AND ELASTICITY OF THREAD. BRILLIANCY AND DURABILITY OF COLORS AND VARIETY OF SHADES
EACH HANK IS SUBDIVIDED INTO FOUR EVEN SKEINS.
10 HANKS TO THE BOX

This ring-label (facsimile above), with the Columbia shepherdess trade mark, is around every hank of genuine Columbia Yarn.

Be sure it is always on the yarn you buy, to get the most satisfactory results from your knitting and crocheting.

Besides the "ring-label," we illustrate the Columbia Carton—showing how the yarn is packed.

The system of packing and inspecting Columbia Yarn is very thorough and very strict. You will find their excellence consistent through the greatest number of packages.

232

Plates from Horstmann's 1917 *Columbia Book of Yarns*. The image of Columbia as a shepherdess was the logo of the non-military side of Horstmann's business which issued this catalog in 1917.

International Exhibition, Philadelphia--1876.

The United States Centennial Commission has examined the report of the Judges, and accepted the following reasons, and decreed an Award in conformity therewith.

PHILADELPHIA, JANUARY 15, 1877.

REPORT ON AWARDS.

Award No.1—Product: DRESS, CARRIAGE & UPHOLSTERY TRIMMINGS.

Name and address of Exhibitor, **WM. H. HORSTMANN & SONS, Philadelphia.**

The undersigned, having examined the product herein described, respectfully recommends the same to the United States Centennial Commission for award, for the following reasons, viz:

For a splendid exhibit of *Dress, Carriage* and *Upholstery Trimmings*, of great excellence and beauty in style, material and execution; also, for a very handsome and complete assortment of *Woolen* and *Mohair Yarns*, known to the trade as *Germantown, Cashmere, Saxony, Shetland* and *Balmoral Yarns*, of brilliant colors, variety of shades, and regularity of spinning.

GUSTAV GEBHARD.
(Signature of the Judge.)

Approval of Group Judges:
ELLIOT C. COWDIN, CHAS. LE BOUTILLIER, CHAS. J. ELLIS, HENRY MITCHELL,
JOHN G. NEESER, JOHN L. HAYES, JOHN D. LANG.

FRANCIS A. WALKER,
Chief of the Bureau of Awards.

Given by authority of the United States Centennial Commission:

A. T. GOSHORN, *Director General.* J. L. CAMPBELL, *Sec'y.* J. R. HAWLEY, *Pres't.*

Award No. 2.—Product: FLAGS, INSIGNIA, EMBLEMS.

The undersigned, having examined the product herein described, respectfully recommends the same to the United States Centennial Commission for award, for the following reasons, viz: *for the excellence of workmanship, variety of exhibit, and correctness in design.*

GEORGE HEWSTON.
(Signature of the Judge.)

Approval of Group Judges:
B. F. BRITTON, KANITZ, DIETZ MONNIN, W. O. LUITHICUM,
E. N. HORSFORD, W. H. CHANDLER, MODEST KITTARY, M. P. EMPEY.

Award No. 3.—Product: SWORDS.

The undersigned, having examined the product herein described, respectfully recommends the same to the United States Centennial Commission for award, for the following reasons, viz: *for elegance of finish and fabrication.*

HENRY L. ABBOTT, Brevet Brig. Gen. U. S. A.
(Signature of the Judge.)

Approval of Group Judges:
W. H. NOBLE, S. C. LYFORD, LUIZ DE SALDANTRA, ALPHONSE LESNE.

A true copy of the record.

FRANCIS A. WALKER,
Chief of the Bureau of Awards.

Approved by authority of the United States Centennial Commission:

A. T. GOSHORN, *Director General.* J. L. CAMPBELL, *Sec'y.* J. R. HAWLEY, *Pres't.*

Catalog showing Centennial Exhibition awards

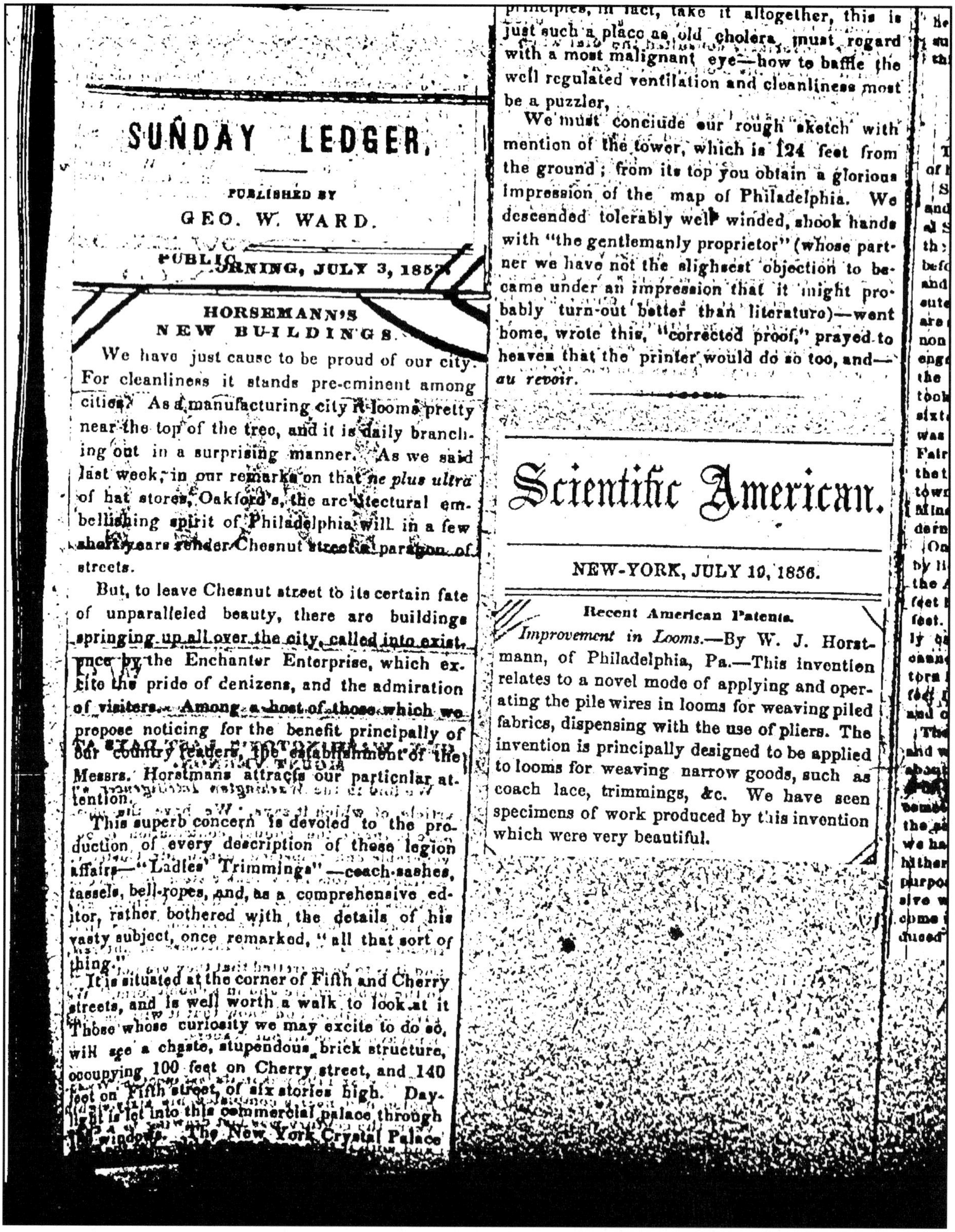

SUNDAY LEDGER.

PUBLISHED BY

GEO. W. WARD.

PUBLI[illegible]RNING, JULY 3, 185[illegible]

HORSEMANN'S NEW BUILDINGS.

We have just cause to be proud of our city. For cleanliness it stands pre-eminent among cities. As a manufacturing city it looms pretty near the top of the tree, and it is daily branching out in a surprising manner. As we said last week, in our remarks on that *ne plus ultra* of hat stores, Oakford's, the architectural embellishing spirit of Philadelphia will in a few short years render Chesnut street a paragon of streets.

But, to leave Chesnut street to its certain fate of unparalleled beauty, there are buildings springing up all over the city, called into existence by the Enchanter Enterprise, which excite the pride of denizens, and the admiration of visiters. Among a host of those which we propose noticing for the benefit principally of our country readers, the establishment of the Messrs. Horstmans attracts our particular attention.

This superb concern is devoted to the production of every description of these legion affairs—"Ladies' Trimmings"—coach-sashes, tassels, bell-ropes, and, as a comprehensive editor, rather bothered with the details of his vasty subject, once remarked, "all that sort of thing."

It is situated at the corner of Fifth and Cherry streets, and is well worth a walk to look at it. Those whose curiosity we may excite to do so, will see a chaste, stupendous brick structure, occupying 100 feet on Cherry street, and 140 feet on Fifth street, of six stories high. Daylight is let into this commercial palace through [illegible] windows. The New York Crystal Palace

[illegible] principles, in fact, take it altogether, this is just such a place as old cholera must regard with a most malignant eye—how to baffle the well regulated ventilation and cleanliness must be a puzzler.

We must conclude our rough sketch with mention of the tower, which is 124 feet from the ground; from its top you obtain a glorious impression of the map of Philadelphia. We descended tolerably well winded, shook hands with "the gentlemanly proprietor" (whose partner we have not the slighest objection to became under an impression that it might probably turn out better than literature)—went home, wrote this, "corrected proof," prayed to heaven that the printer would do so too, and—*au revoir*.

Scientific American.

NEW-YORK, JULY 19, 1856.

Recent American Patents.

Improvement in Looms.—By W. J. Horstmann, of Philadelphia, Pa.—This invention relates to a novel mode of applying and operating the pile wires in looms for weaving piled fabrics, dispensing with the use of pliers. The invention is principally designed to be applied to looms for weaving narrow goods, such as coach lace, trimmings, &c. We have seen specimens of work produced by this invention which were very beautiful.

William J. Horstmann thoroughly understood the manufacture of trimmings and invented several improvements in the process. (*Scientific American,* July 7, 1856, from the Horstmann scrapbooks)

Index